Essential Calculus

with Motivational Problems

Bob Smither, Ph.D.

C-C-I Press

Friendswood, Texas

Essential Calculus
by Bob Smither, Ph.D.
Copyright ©2010 Bob Smither, Ph.D.
Published by C-C-I Press, Friendswood, Texas
Created using Open Source Tools:
___ LyX, XFig, and GnuPlot on Linux
Printed in the United States of America
Printing History:
___ First Edition: February 2010

Bob Smither is president of Circuit Concepts, Inc., an electronic consulting company in Friendswood, Texas. He received his Ph.D. in electrical engineering from the University of Houston in 1978 and taught electrical engineering there for 14 years.

Cataloging Data
Smither, Bob.
Essential Calculus / Bob Smither, Ph.D.
ISBN 9781450545846
1. Calculus
QA303
517

Dedicated to David

Contents

I. Background and Motivation   1

1. Background   3
1.1. Functions of a Single Variable   3
1.2. Functions of Multiple Variables   5
1.3. Taking Limits   7
1.4. Limits, the Value of e, and Compound Interest   9
1.5. Taking a Limit Numerically   11
1.6. Multiple Angle Formulas   14

2. Motivation   17
2.1. Slope of a Curve   17
2.2. The Capacitor   19
2.3. Age of the Earth[5]   21
2.4. Designing Pastures   21
2.5. Maximizing Profit   22
2.6. Euler's Formula   22
2.7. Maximum Package Volume   24
2.8. What is the value of $0/0$?   24
2.9. What is the value of 1^∞?   25
2.10. Acceleration, Speed, Distance   25
2.11. The Longest Shot   25
2.12. Distance Traveled   26
2.13. Area of a Circle   27

Contents

2.14. Potential Energy of Gravity 27

II. Differential Calculus   33

3. The Derivative   35

3.1. The Derivative - the Slope of a Curve 35
3.2. Motivational Problem 2.1 - Solution 37
3.3. Linearity of Differentiation 38
3.4. Derivative of Sin(x) and Cos(x) 40
3.5. Derivative of a Polynomial 41
3.6. Deriving Derivatives using Series Expansions 44
    3.6.1. Derivative of Sin(x) using Series Expansion 44
    3.6.2. Derivative of Cos(x) using Series Expansion 45
    3.6.3. Derivative of the Exponential Function using Series Expansion 46
3.7. Derivative of an Exponential 46
3.8. Motivational Problem 2.2 - Solution 49
3.9. Motivational Problem 2.3 - Solution 51
3.10. Derivative of the Logarithm 52
3.11. Minimum and Maximum Values 54
3.12. Motivational Problem 2.4 - Solution 56
3.13. An Even Better Pasture? 58
3.14. Motivational Problem 2.5 - Solution 58
3.15. The Product Rule 60
3.16. Motivational Problem 2.6 - Solution 62
3.17. Euler's Identity 63
3.18. The Quotient Rule 63
3.19. Motivational Problem 2.7 - Solution 65
3.20. The Chain Rule 67

3.21. Derivative of a Function with Scaled Independent Variable 69

3.22. L'Hôpital's Rule for Indeterminate Limit of Type 0/0 71

3.23. Motivational Problem 2.8 - Solution 73

3.24. L'Hôpital's Rule for Indeterminate Limit of the Type ∞/∞ 74

3.25. Motivational Problem 2.9 - Solution 75

III. Integral Calculus 77

4. The Integral 79

4.1. The Integral - the Area under a Curve . . . 79

    4.1.1. Area Using the Riemann Sum 79

    4.1.2. A Better Way - The Fundamental Theorem of Calculus 81

4.2. Two Types of Integrals 85

    4.2.1. Definite Integral 85

    4.2.2. Indefinite Integral 87

4.3. Motivational Problem 2.10 - Solution 87

4.4. Motivational Problem 2.11 Solution 91

4.5. Motivational Problem 2.12 - Solution 97

    4.5.1. Using Riemann Sum 99

    4.5.2. Using the Fundamental Theorem of Calculus 101

4.6. Motivational Problem 2.13 - Solution 103

4.7. Simple Integrals 106

4.8. Motivational Problem 2.14 - Solution 107

4.9. Integration using Substitution 109

    4.9.1. Example: $\int(a*x+b)dx$ 109

    4.9.2. Example: $\int \sin(a*x+b)dx$ 110

    4.9.3. Example: $\int_0^\pi \sin(a*x+b)dx$ 111

Contents

4.9.4. Example: $\int \frac{dx}{(a+x)}$ 111

4.9.5. Example: $\int \frac{dx}{(a+x)^2}$ 112

4.9.6. Example: $\int \ln(a * x + b) dx$ 112

4.10. Integration by Parts 113

4.10.1. Example: $\int x * \sin(a * x) dx$ 114

4.10.2. Example: $\int \log_b(a * x) dx$ 115

A. Collection of Functions   **117**

Index   **119**

Bibliography   **121**

List of Figures

1.1. Limit as h approaches 0.0 12

2.1. Straight Line Plot 18
2.2. X Squared Curve 19
2.3. Discharging a Capacitor 20
2.4. Price versus Widgets - Demand Curve . . . 23
2.5. Cost versus Widgets - Cost Curve 23
2.6. The Longest Shot 26
2.7. Variable Speed vs. Time 27
2.8. Circle of Radius r 28
2.9. Gravitational Potential Energy 29
2.10. Gravitational Potential Energy 30

3.1. Exponential Function 48
3.2. Minimum and Maximum of a Curve 54
3.3. $y(x) = x^3, y'/dx = 3 * x^2, y'' = 6 * x$ 55
3.4. Approximating y(x) near x=a 72

4.1. Dividing the Time axis into Segments 80
4.2. Area Under the Curve from a to x. 82
4.3. Initial Velocity Vector 92
4.4. x_F versus θ 96
4.5. The Longest Shot 97
4.6. Speed vs. Time Plot 98
4.7. Distance vs. Time Plot - Constant Speed . . 98
4.8. Speed vs. Time Plot - Constant Speed . . . 99

List of Figures

4.9. Speed vs. Time 100

4.10. Area of a Circle 104

4.11. Area of Circle as Area Under a Curve 105

List of Tables

1.1. Functions of a Single Variable   4
1.2. Limit of $3*\sin(x)/x$ 8
1.3. Limit as h approaches 0   11

3.1. Estimating the Slope of the Line   36
3.2. Binomial Coefficients   42
3.3. Polynomial Derivatives   43
3.4. Min/Max and Inflection Points   56

List of Tables

Preface

This book is titled <u>Essential Calculus</u> for a reason. It is written to present the basics of calculus in an easy to understand way. If you want to be exposed to an overview of calculus, with enough depth to provide understanding and the ability to solve real world problems, I hope that this book will satisfy that urge.

Calculus is a subject which can change the way that one looks at the world. It is not difficult and can be eye-opening and exciting to learn. There are many "AHA!" moments available to the careful student of calculus.

I hope that this book will allow me to share some of those moments with you.

Why did I write this book? My son is 22 and for various reasons still has not taken a calculus course. My background is in Electrical Engineering and I have anticipated teaching my son calculus ever since the two of us would jog together doing multiplication tables when he was 7.

My son has taken a calculus pre-requisite class in college, but is focused on other areas and still has not taken the course. Perhaps he will find this book useful when he finally takes Calculus.

Bob Smither
February, 2010

Part I.

Background and Motivation

1. Background

1.1. Functions of a Single Variable

The idea of a function is important to the study of calculus. A function can be represented as

$$y = f(x)$$

which is read as "y is a function of x." In order to be a function, for each allowed value of x the function f must produce *one and only one* value of y. The set of all allowed values of x is called the domain of the function and the set of all resulting values of y is called the range or co-domain of the function f. A few examples of functions of a single variable are shown in Table 1.1.

In table 1.1 the symbol $\mathbb{R}$ is used for the real line (all values from $-\infty$ to $+\infty$) and NAF means "not a function." Can you see why $y = \arccos(x)$ and $y = \arctan(x)$ are not functions (hint - plot the relationship)? Actually the inverse trigonometric functions can be "cleaned up" if we only consider what is known as the "principle value" range of the function. As an example, if we restrict the range of the $\arccos(x)$ to $0 \leq \arccos(x) \leq \pi$ then the domain of this function is $-1 \leq x \leq +1$. Can you determine the range of values for the $\arctan(x)$ which would allow us to call this a function?

Function	Domain (x)	Range (y)
$y = 1$	$\mathbb{R}$	1
$y = x$	$\mathbb{R}$	$\mathbb{R}$
$y = x^2$	$\mathbb{R}$	$y \geq 0$
$y = x^3$	$\mathbb{R}$	$\mathbb{R}$
$y = \cos(x)$	$\mathbb{R}$	$\mid y \mid \leq 1$
$y = \arccos(x)$	NAF	-
$y = \tan(x)$	$\mathbb{R}$	$\mathbb{R}$
$y = \arctan(x)$	NAF	-
$y = \ln(x)$	$x \geq 0$	$\mathbb{R}$
$y = \exp(x)$	$\mathbb{R}$	$y \geq 0$

Table 1.1.: Functions of a Single Variable

Worth Remembering:
The expression

$$y = f(x)$$

defines y as a function of x only if for every allowed value of x there is one and only one resulting value of y.

Domain of f: The set of allowed values of x.

Range of f: The set of resulting values of y.

1.2. Functions of Multiple Variables

It is also possible to talk about functions of more than one variable. The lift of an airplane wing, for example, might be a function of air speed over the wing, the density of the air, and the angle of attack of the wing. The elevation of the land above sea level is a function of both the latitude and longitude. A common notation used for a function of two variables is:

$$z = f(x, y)$$

and would be read as "z is a function of x and y."

A simple example of a function of two variables is the equation for a plane in three dimensions:

$$a * x + b * y + c * z = d.$$

Writing this as a function gives:

1. Background

$$f(x, y) = z = A * x + B * y + D$$

where the constants A, B, and D can be written in terms of a, b, c, and d as:

$$A = -a/c$$

$$B = -b/c$$

$$D = d/c$$

Just as with single variable functions we can talk about the domain and range of a function of multiple variables. The domain of x and y is $\mathbb{R}$, and the range of z is also $\mathbb{R}$ (unless both A and B are 0).

As another example, consider

$$f(x, y) = \sqrt{x} + \sqrt{y}.$$

Since we can't take the square root of a negative number the domain of the function must be

$$x \geq 0; \; y \geq 0$$

The resulting range of the function is $\mathbb{R}$.

1.3. Taking Limits

Taking limits is central to calculus. What does it mean to take a limit of a function? The notation used is

$$\lim |_{x \to a} f(x) = b$$

and is read as "the limit of $f(x)$ as x goes to a equals b." In most functions it is not difficult to find the limit. For example

$$\lim |_{x \to \pi} \cos(x) = -1$$

since we can just plug in the limit value (π) into the function ($\cos(x)$) to get the limit.

Formally the limit is defined as follows:

> We understand the statement
>
> $$lim |_{x \to a} f(x) = b$$
>
> as meaning that for any positive number ε we can find a positive number δ such that whenever
>
> $$| x - a | < \delta$$
>
> we have
>
> $$| f(x) - b | < \varepsilon$$

In words, we can say that the value of the function $f(x)$ gets arbitrarily close (but not necessarily equal) to the limiting

δ	x	ϵ	$3*\sin(x)/x$
1.0	1.0	0.475...	2.524...
0.5	0.5	0.1234...	2.8765...
0.05	0.05	0.00124...	2.99875...
0.005	0.005	0.000012...	2.999987...

Table 1.2.: Limit of $3*\sin(x)/x$

value b when the value of x gets arbitrarily close to the limit value a.

For some functions the limit may not be so obvious - consider:

$$lim\ |_{x\to0}\ 3*\sin(x)/x.$$

If we just plug in the limit value (0) into the function $3*\sin(x)/x$ we get $0/0$ which has no meaning.

If we try some small values of x it appears that the function $3*\sin(x)/x$ is approaching the limit value of 3.0. Using this limit value and the definition above we can explore what this limit process means. Table 1.2 shows the values of δ, x, ε, and $3*\sin(x)/x$ for smaller and smaller values of ϵ.

It is clear that the function $3*\sin(x)/x$ is tending to 3.0 as x tends to 0.0. If we continue this numerical process the required precision will overwhelm any physical calculator. For a detailed discussion of this precision problem, see Section 1.5.

Using calculus we will develop a better method of handling such awkward limits.

1.4. Limits, the Value of e, and Compound Interest

We will need the following result later in this book. Consider normal, annual compound interest. Starting with an initial deposited principle of P_o compounded at an annual rate r, after a year, we have

$$P_1 = (1 + r) * P_0.$$

For example, an initial deposit of $1000 will grow to

$$P_1 = \$1000 * (1 + .08) = \$1080$$

after a year if the annual rate is 8%. If the deposit is left for multiple years, the principle, compounded annually, will grow to

$$P_n = P_0 * (1 + r)^n$$

after n years.

Now, what happens if the compounding is done say twice a year? To be consistent, let's use r as the annual rate. If we compound twice in a year, this means that we will get interest at an $r/2$ rate the first half of the year and at the same rate the second half of the year. The cool part about twice a year compounding is that the interest earned in the first half of the year gets compounded along with the principal in the second half. In equation form

$$P_{1/2} = P_0 * (1 + r/2)$$

where $P_{1/2}$ is the principal plus interest after a half year. Then

$$P_1 = P_{1/2} * (1 + r/2)$$

9

covers the second half of the year so that

$$P_1 = P_0 * (1 + r/2)^2.$$

Since the interest gets compounded for the second half of the year, we expect to have more at the end of the year and sure enough, for the same 8% annual rate, six month compounding results in

$$P_1 = \$1000.00 * (1 + .04)^2 = \$1081.60.$$

Not much improvement, but better than annual compounding. This exercise suggests the following question - what if we compound many times during the year? In fact, what if we compound continuously (an infinite number of times) during the year? This new situation can be expressed as a limit:

$$P_1 = lim \, |_{n \to \infty} \, P_0 * (1 + r/n)^n.$$

Taking a limit means letting the limit variable, in this case n, approach the limiting value, in this case ∞. Here we can get a feel for how the limit works by letting n get large. Before we actually investigate this limit, lets rearrange things a bit. Following [11], look at just the part involved in the limit and call it L:

$$L = \lim \, |_{n \to \infty} \, (1 + r/n)^n.$$

Now let $n = r/h$ and note that the limit $n \to \infty$ gets replaced with $h \to 0$ so that we have

$$\lim \, |_{h \to 0} \, (1 + h)^{r/h}$$

or

h	$(1+h)^{1/h}$
1.00000	2.0000
0.10000	2.5937
0.01000	2.7048
0.00100	2.7169
0.00010	2.7181
0.00001	2.7183

Table 1.3.: Limit as h approaches 0

$$\lim |_{h \to 0} \left((1+h)^{1/h} \right)^r$$

(recall that $x^{a*b} = (x^a)^b$).

Finally, consider only the term that depends on the limit variable h:

$$e = \lim |_{h \to 0} (1+h)^{1/h}.$$

As implied by the notation, this limit in fact equals the transcendental number e, the base of the natural logarithms. This result can be readily demonstrated numerically as shown in Table 1.3.

1.5. Taking a Limit Numerically

If we continue the process outlined in Table 1.3 we run into a problem - this numerical process will eventually break down.

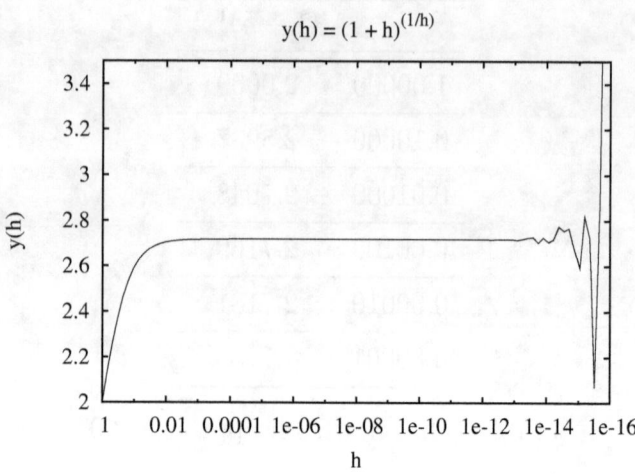

Figure 1.1.: Limit as h approaches 0.0

Using the GnuPlot [2] program we can explore just how close we can get to e as we make h smaller and smaller. Plot 1.1 shows the value of $(1 + h)^{1/h}$ plotted against the value of h.

With $h = 1$, the function $(1+h)^{1/h}$ is 2.0. As h gets smaller and smaller, say in the range of .000001 to .0000000001 $(1 * 10^{-6}$ to $1 * 10^{-10})$, the function $(1 + h)^{1/h}$ appears to approach a limiting value of 2.71828... . As h gets smaller yet, the numerical calculation falls apart due to the finite precision of the underlying calculations. This should come as no surprise - look at the calculation we are trying to do at, for example, $h = 1 * 10^{-10}$. Writing this out we have

$$(1 + 1 * 10^{-10})^{1*10^{10}}$$

so the first thing we need to do is evaluate

$$(1 + 0.0000000001) * (1 + 0.00000000001)$$

or

$$1 + 0.0000000002 + 0.00000000000000000001.$$

Do you see the problem yet? We need to represent a number to 20 decimal places and this is just the first multiplication out of $1 * 10^{10}$ such multiplications! Each additional $(1 + 0.0000000001)$ term in our product adds 10 digits to the precision required. Modern day high precision (high resolution) computers designed for advanced scientific calculations can represent numbers to perhaps 32 decimal digits of precision (so called Quad precision numbers, defined in the IEEE Standard for Floating Point Arithmetic, IEEE 754-2008), so these sophisticated machines are incapable of representing even the third product of our series (which would need 40 decimal digits of precision). There is a technique known as Arbitrary-Precision Arithmetic, but even this is limited by the (necessarily) finite amount of computer memory that can be used for a calculation.

It seems that the numerical approach to taking a limit is, well, limited!

The approximate value of e is 2.71828182845904523...

> Worth Remembering:
> The value of e
>
> $$e \simeq 2.7183$$

Getting back to our interest rate problem we now have

$$L = e^r = \exp(r)$$

so that

$$P_1 = lim \mid_{n \to \infty} P_0 * (1 + r/n)^n$$

$$P_1 = P_0 * \exp(r).$$

For our example of $r = 8\%$, if we compound continuously during the year we end up with

$$P_1 = \$1000 * \exp(.08) \simeq \$1083.29.$$

1.6. Multiple Angle Formulas

We will use some of the multiple angle formulas later in this book, so let's spend a few minutes obtaining them here. We will use Euler's Formula [14] to derive the trigonometric relationships. Later, after we have a little calculus under our belts, we will demonstrate the validity of Euler's Formula - here we will just take it as proven.

Euler's Formula states that

$$\exp(j * \theta) = \cos(\theta) + j * \sin(\theta) \qquad (1.1)$$

where j is the imaginary number $\sqrt{-1}$.

Before looking at multiple angle formulas, let's quickly derive a useful relationship, namely:

$$1 = \cos^2(\theta) + \sin^2(\theta). \qquad (1.2)$$

From equation 1.1, take the complex conjugate of both sides:

$$\exp(-j * \theta) = \cos(\theta) - j * \sin(\theta) \qquad (1.3)$$

Now, multiplying equation 1.1 by equation 1.3 we get:

$$\exp(j * \theta) * \exp(-j * \theta) = \exp(0) = 1 = \cos^2(\theta) + \sin^2(\theta)$$

proving equation 1.2.

Returning to Euler's Formula, we can write

$$\exp(j * (\theta + \phi)) = \cos(\theta + \phi) + j * \sin(\theta + \phi)$$
$$= \exp(j * \theta) * \exp(j * \phi)$$
$$= (\cos(\theta) + j * \sin(\theta)) * (\cos(\phi) + j * \sin(\phi))$$
$$= \cos(\theta) * \cos(\phi) - \sin(\theta) * \sin(\phi) +$$

$$j * [\sin(\theta) * \cos(\phi) * \cos(\theta) * \sin(\phi)]$$

Look at the right hand side of the first and last equations in this set - we can equate the real parts and the imaginary parts to get

$$\cos(\theta + \phi) = \cos(\theta) * \cos(\phi) - \sin(\theta) * \sin(\phi) \qquad (1.4)$$

and

$$\sin(\theta + \phi) = \sin(\theta) * \cos(\phi) + \cos(\theta) * \sin(\phi). \qquad (1.5)$$

These are two of the multiple angle formulas from trigonometry.

1. Background

Additional multiple angle formulas result from letting $\theta = \phi$ in equations 1.4 and 1.5. From equation 1.4:

$$\cos(2 * \theta) = \cos^2(\theta) - \sin^2(\theta)$$

$$= [1 - \sin^2(\theta)] - \sin^2(\theta)$$

(using equation 1.2) from which

$$\sin^2(\theta) = \frac{1}{2} * [1 - \cos(2 * \theta)]. \tag{1.6}$$

Similarly, it is easy to show that

$$\cos^2(\theta) = \frac{1}{2} * [1 + \cos(2 * \theta)]. \tag{1.7}$$

2. Motivation

This chapter presents several motivational problems that can be solved using the ideas developed in this book. It is hoped that by presenting these problems early in the book you will be motivated to continue so you can learn and appreciate how real world problems can be solved using calculus. The solutions to each of the motivational problems will be presented later in the book after the material required for their solution has been covered.

Don't be surprised if you do not understand all of the ideas presented in these problems! You are not expected to. Complete explanations and solutions will be presented after we have covered the calculus ideas needed to solve each Motivational Problem.

2.1. Slope of a Curve

We are all familiar with the slope - intercept representation of a straight line [3]:

$$y = m * x + b$$

where m is the slope of the line and b is the y axis intercept (the value of y when x = 0). The slope, m, is the ratio of

2. Motivation

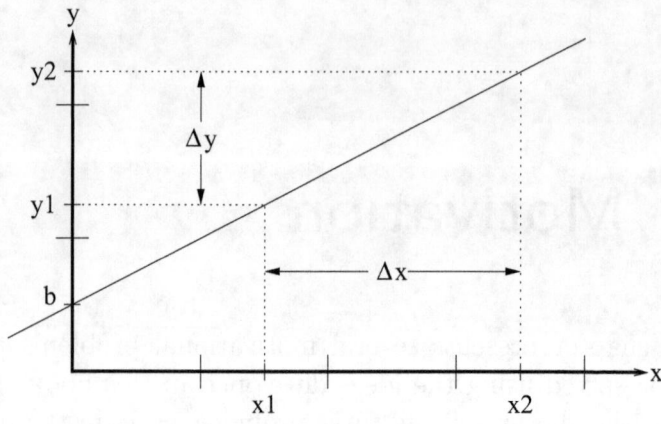

Figure 2.1.: Straight Line Plot

the *change* in the y direction divided by the *change* in the x direction as one moves along the line. In other words:

$$m = \frac{y_2 - y_1}{x_2 - x_1} = \frac{\Delta y}{\Delta x}$$

where the Δ is measured over some convenient length. See Figure 2.1 for the details.

Measuring the slope of a straight line is pretty straightforward, but what about the slope of a *curve*? Looking at Figure 2.2 it is clear that the slope of the line changes as x changes - in fact the slope can be seen to be a function of x.

Motivational Problem 2.1 is to find the function of x that is the slope of the curve shown in Figure 2.2 for any value of x.

Solution: Section 3.2 on page 37.

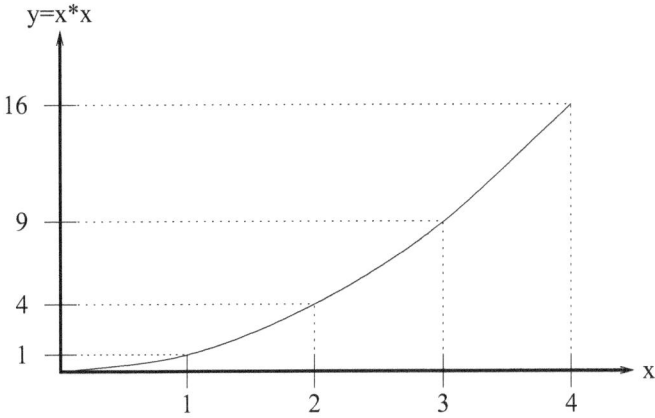

Figure 2.2.: X Squared Curve

2.2. The Capacitor

A capacitor is a two terminal electrical element that is used to store electrical charge. Capacitors are used in many industrial applications and in just about all electronic devices. The basic operation of a capacitor is described by

$$Q = C * V$$

where Q is the charge stored on the capacitor, V is the voltage measured across the capacitor's terminals, and C is a measure of the "capacitance" of the capacitor. The units are: Coulombs for Q, Volts for V, and Farads for C.

The time rate of charge movement in a conductor is current. Denoting current as I and noting that current in the direction defined in Figure 2.3 will decrease the charge we have

$$I = -dQ/dt.$$

Current is measured in Amperes. The flow of charge, that is current, through a resistance (R) causes a voltage difference

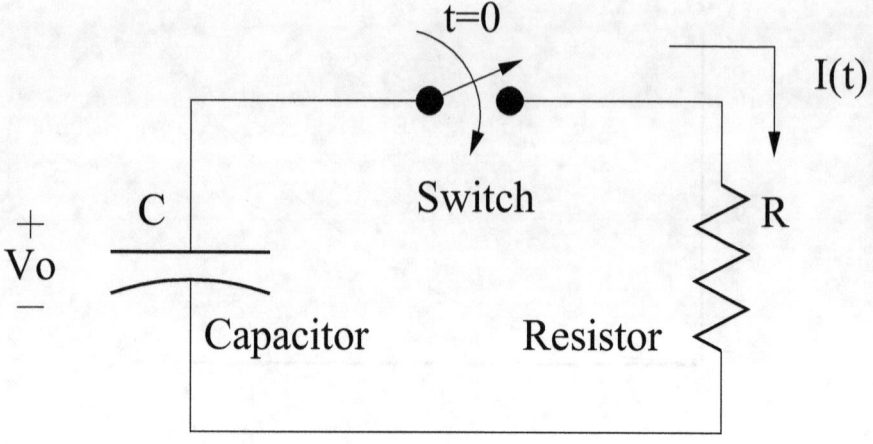

Figure 2.3.: Discharging a Capacitor

to appear between the ends of the resistor. The equation expressing this fact is known as Ohm's Law:

$$V = I * R$$

where V is the voltage drop between the ends of the resistor. Resistance is measured in Ohms.

Figure 2.3 illustrates an experiment in which a capacitor of capacitance C, originally charged to V_o Volts, is connected across a resistor of resistance R at time $t = 0$ by closing the switch. Before time $t = 0$ the switch is open, the voltage on the capacitor is V_o, and the current $I(t) = 0$.

Motivational Problem 2.2 is to find the equation for the current $I(t)$ that flows through the resistor R for time ≥ 0.

Solution: Section 3.8 on page 49.

2.3. Age of the Earth[5]

The atoms of certain elements spontaneously decay through emission of radiation or particles. While each decay event is random, given a large number of radioactive atoms the average rate is predictable. Clearly the more such atoms there are the more will decay, so we expect the rate at which radioactive atoms are lost to be proportional the number of such atoms. Remember that the derivative relates to a rate, so it is natural to write

$$dN/dt = -\lambda * N. \qquad (2.1)$$

(see Section 3.1 for an introduction to the notation used here) The constant of proportionality, λ, is positive so the rate, dN/dt, is a negative number implying that the number of radioactive atoms decreases as time progresses.

It turns out that two isotopes of uranium, ^{235}U and ^{238}U, decay at different rates. It also develops, as best we know, that the two isotopes should have been created in approximately equal quantities and their abundance should have been approximately equal around the time the earth was forming. The current ratio of ^{238}U to ^{235}U on earth is 137.8. The λ for ^{235}U is 9.80E-10 and the λ for ^{238}U is 1.55E-10.

Motivational Problem 2.3 is to estimate the age of the earth from the above information.

Solution: Section 3.9 on page 51.

2.4. Designing Pastures

Pastures are often defined by fences, and in designing a pasture for livestock an important consideration is the cost.

Let's assume that fencing costs a fixed amount per foot, say $F/ft.$, and that a rancher has a budgeted amount, say B, that he is prepared to spend on fencing a new, rectangular, pasture. Motivational Problem 2.4 is: How should the pasture be designed so that the rancher gets the largest possible fenced area for his expenditure of B?

Solution: Section 3.12 on page 56.

2.5. Maximizing Profit

Let's assume we want to manufacture and sell some Widgets (W). The market for Widgets has a demand curve as shown in Figure 2.4 - that is, we can only sell more Widgets if we are willing to lower the price. The Figure tells us that we will sell no Widgets at $10.00/W, but can increase our sales by 100W each time we lower the price by $2.50. The demand curve has a slope of -$0.025/W.

There is also a cost curve associated with producing and selling Widgets as shown in Figure 2.5. After a fixed cost of $500, each Widget costs us $2.50 to manufacture.

Motivational Problem 2.5 is to determine the number of Widgets we should make in order to maximize our profit.

Solution: Section 3.14 on page 58.

2.6. Euler's Formula

Motivational Problem 2.6 is to demonstrate the truth of Euler's Formula, which is:

$$\exp(j * \theta) = \cos(\theta) + j * \sin(\theta).$$

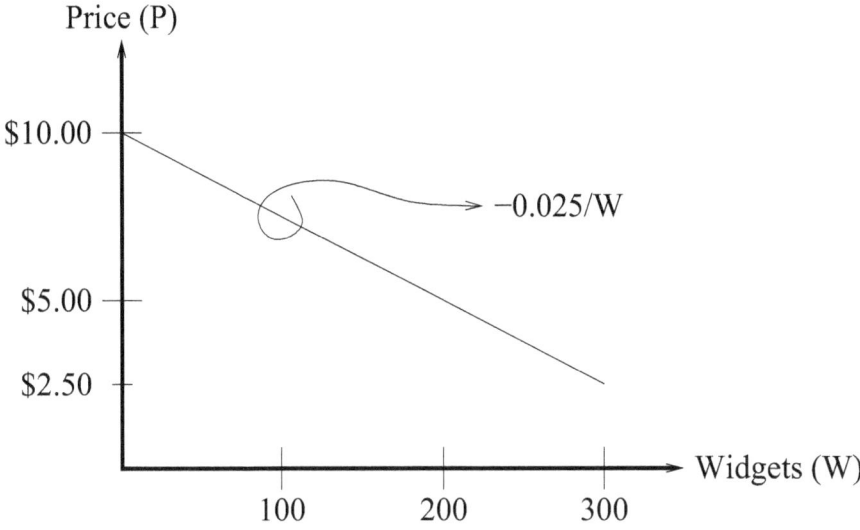

Figure 2.4.: Price versus Widgets - Demand Curve

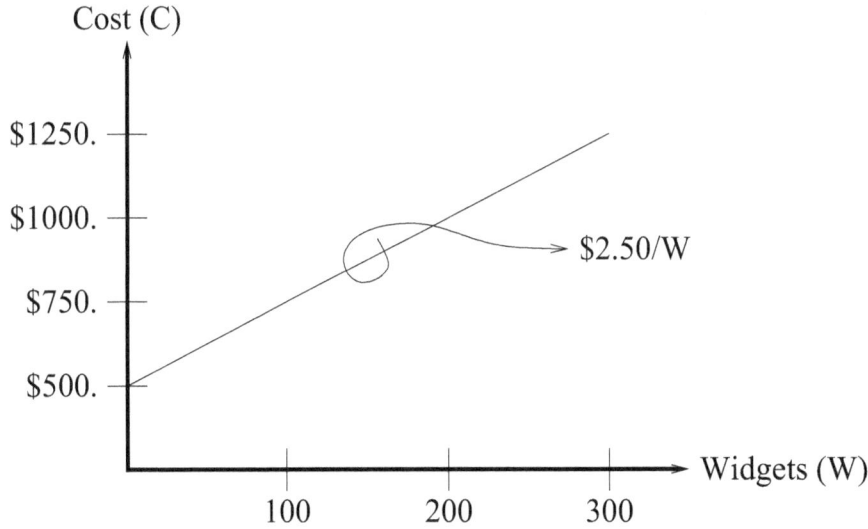

Figure 2.5.: Cost versus Widgets - Cost Curve

Solution: Section 3.16 on page 62

2.7. Maximum Package Volume

In the United States the UPS company restricts packages to a combined girth plus length of 165 inches with a maximum allowed length of 108 inches. Length is defined as the length of the longest side of the rectangular package. If we want to ship a large quantity of small items, we want to maximize the volume of the package that we use for shipments.

Motivational Problem 2.7 is to find the dimensions of the largest volume package that UPS will ship for us.

Solution: Section 3.19 on page 65.

2.8. What is the value of $0/0$?

We sometimes run into expressions such as

$$f(x) = 3 * \sin(x)/x$$

and we need to know the value of $f(x)$ when x goes to 0.0. In other words we need to find

$$lim \mid_{x \to 0} 3 * \frac{\sin(x)}{x}.$$

Substituting the limit value of 0 gives $\frac{0}{0}$ - not very useful. Motivational Problem 2.8 is to find the value of the above limit.

Solution: Section 3.23 on page 73.

2.9. What is the value of 1^∞?

In Section 1.4 we developed this limit:

$$\lim \big|_{h\to 0} (1+h)^{1/h}$$

which we declared was equal to the transcendental number
e. Substituting the limit value of 0 results in 1^∞ - not
easily understood. Motivational Problem 2.9 is to show
that indeed this limit equals e.

Solution: Section 3.25 on page 75.

2.10. Acceleration, Speed, Distance

If an object is accelerated at a constant acceleration a (with
units of $m/sec/sec$) it's speed will increase by $a(m/sec)$ for
each second that it is accelerated. For example, an object
dropped from a tower on earth experiences an acceleration
of $9.8m/sec/sec$ so that three seconds after being dropped it
will be moving at $29.4m/sec$ (ignoring the effect of air drag
and assuming that the object has not hit the ground!).

Motivational Problem 2.10 is to obtain the equation for the
speed and the distance traveled by an object subject to a
constant acceleration a.

Solution: Section 4.3 on page 87.

2.11. The Longest Shot

In firing a cannon one of the variables that can be controlled
is the angle that the cannon makes with the horizontal. Fig-

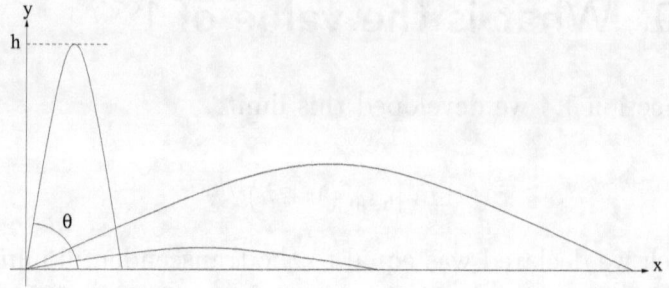

Figure 2.6.: The Longest Shot
The Longest Shot

ure 4.5 illustrates the situation being considered - a cannon is fired on a flat plane. Each time the cannon is fired, the cannon ball is the same size and weight and the initial speed of the ball is the same. Motivational Problem 2.11 is, ignoring wind resistance and assuming that gravity is constant, find the angle Θ that results in the longest possible shot from the cannon.

Solution: Section 4.4 on page 91.

2.12. Distance Traveled

Motivational Problem 2.12 is to find the distance traveled after 60 seconds if the speed versus time curve is as shown in Figure 2.7.

Solution: Section 4.5 on page 97

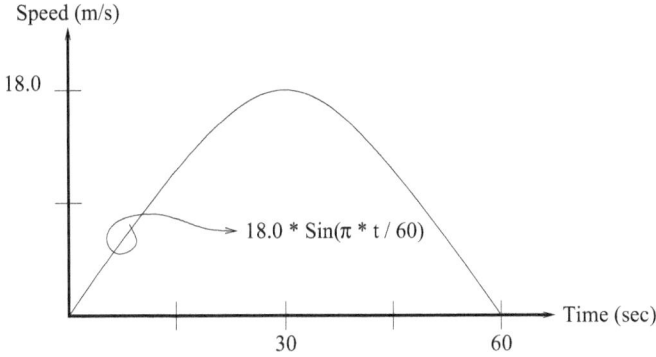

Figure 2.7.: Variable Speed vs. Time

2.13. Area of a Circle

The area of a square is easy, as we saw in Figure 4.8, but how about the area of a circle? Figure 2.8 shows a circle of radius R. We know from earlier math classes that the area of the circle is $\pi * R^2$, but how can this be demonstrated?

Motivational Problem 2.13 is to show that the area of a circle with a radius of R is

$$A = \pi * R^2.$$

The Figure gives a hint of how we will go about solving this problem.

Solution: Section 4.6 on page 103.

2.14. Potential Energy of Gravity

Gravity near the earth's surface exerts a constant acceleration on mass. The force of gravity, $F = m * g$ (Force (F)

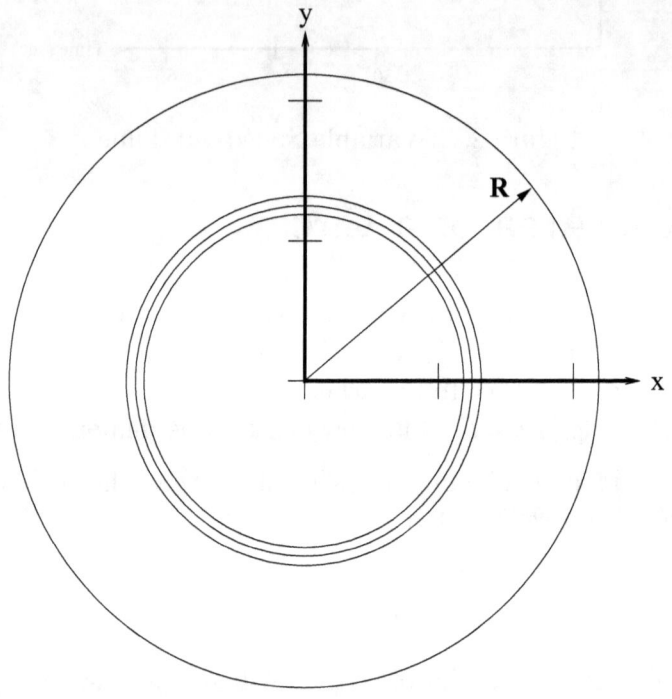

Figure 2.8.: Circle of Radius r

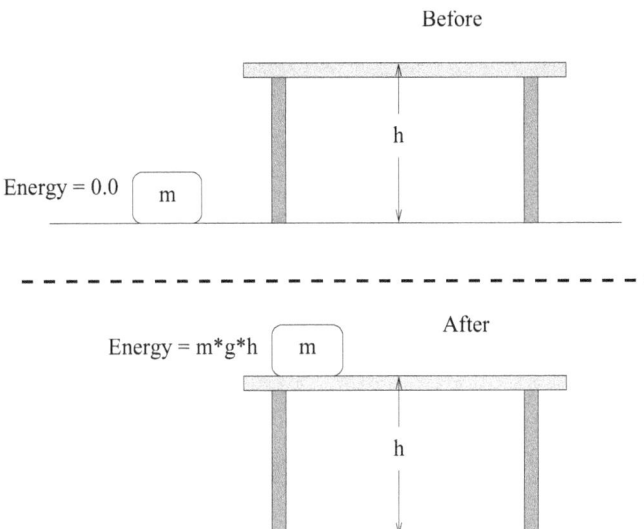

Figure 2.9.: Gravitational Potential Energy

equals mass (m) times acceleration (g)), is what we experience as our weight. To a physicist, work is equal to force times the distance that the force acts over. If we lift a mass against the force of gravity we do work on the mass and give it potential energy. In the case of constant acceleration, as seen near the earth's surface, the force is constant and the potential energy is given by $m * g * h$ where g is the (constant) acceleration of gravity and h is the height that the object is lifted. See Figure 2.9.

What happens when the force of gravity is not constant? As we go away from the earth's surface, the acceleration of gravity reduces. The simple analysis above that takes the acceleration of gravity g to be constant is no longer right. Figure 2.10 illustrates the new situation.

Motivational Problem 2.14 is to find an expression for the work done on a mass m as it is removed from the vicinity of

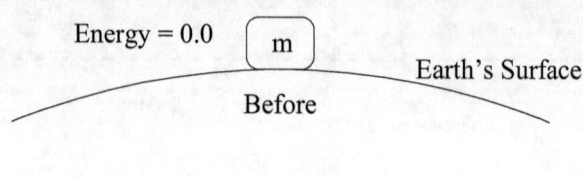

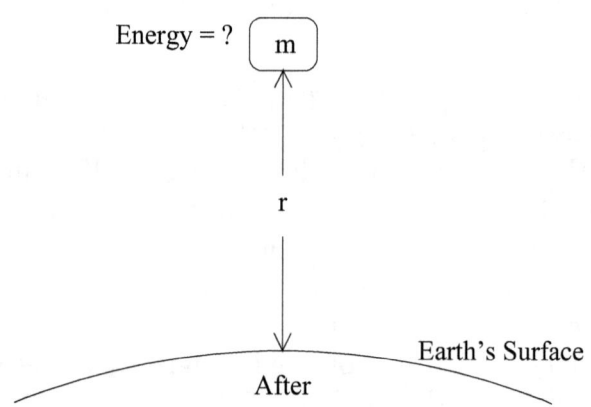

Figure 2.10.: Gravitational Potential Energy

the earth, or what amounts to the same thing, find the potential energy due to gravity for a mass at an arbitrary distance h measured from the earth's surface. Energy is always defined relative to an arbitrary reference, so you can assume that the energy of the mass resting on the earth's surface is zero. Use the resulting expression to find the speed needed by a mass m at the earth's surface to completely escape the pull of earth's gravity (complicating issues like the drag of the earth's atmosphere can be ignored!).

Solution: Section 4.8 on page 107.

2. Motivation

Part II.

Differential Calculus

3. The Derivative

This chapter presents the idea of the derivative as the slope of a curve, and also demonstrates some important properties of the derivative. The properties are derived using the generic definition of a derivative as a slope calculated over smaller and smaller intervals, that is we define the derivative of function $y(x)$ with respect to the variable x as

$$derivative\ of\ y(x)\ with\ respect\ to\ x = slope\ of\ y(x)\ at\ x$$

$$= lim\ |_{\Delta \to 0}\ \frac{y(x + \Delta) - y(x)}{\Delta}.$$

3.1. The Derivative - the Slope of a Curve

Say we want to find the slope of the curve in Figure 2.2 on page 19 at the point x = 2. The brute force method might go something like this:

- Let x_1 and x_2 be 2 and $2 + \Delta$, respectively.

- Calculate y_1 (value of y for $x = 2$) and y_2 (value of y for $x = 2 + \Delta$).

Δ	Slope
0.5	4.5000000...
0.1	4.1000000...
0.01	4.0100000...
0.001	4.0010000...
0.0001	4.0001000...
0.00001	4.0000100...
0.000001	4.0000010...
0.0000001	4.0000001...

Table 3.1.: Estimating the Slope of the Line

- Estimate the slope as

$$\frac{y_2 - y_1}{x_2 - x_1} = \frac{y_2 - y_1}{\Delta}$$

.

The choice for the value of Δ is not obvious. Clearly it should be *small*, whatever that means in this context. If we take $\Delta = 1$ (clearly not small) we get $x_1 = 2$, $x_2 = 2 + \Delta = 3$, $y_1 = 2^2 = 4$, and $y_2 = 3^2 = 9$ so that:

$$m = \frac{9 - 4}{1} = 5$$

As we will see in a minute, this is seriously wrong.

In Table 3.1 we do the above calculation for different, smaller, values of Δ:

It is clear that the slope is approaching 4.0. Using arithmetic to estimate the slope can provide us with some insight, but at some point, as we saw in Section 1.4, the precision of the calculator will be insufficient. This arithmetical approach is prone to error and is cumbersome. There must be a better way!

Let's look at this problem from another angle. We can express the slope analytically as a function of Δ, leaving the value of Δ undetermined for a moment:

$$m \mid_{x=2} = \frac{y_2 - y_1}{x_2 - x_1} = \frac{(2 + \Delta)^2 - (2)^2}{\Delta} = \frac{(4 + 4 * \Delta + \Delta^2) - 4}{\Delta}$$

which gives

$$m \mid_{x=2} = 4 + \Delta$$

We know that Δ should be small to get an accurate answer. We can simply let $\Delta \to 0$ and we have the slope as exactly 4.

This same approach can be further generalized, and we can use the result to solve Motivational Problem 2.1.

3.2. Motivational Problem 2.1 - Solution

Write the expression for the slope in terms of both Δ and an arbitrary value of x:

$$m = \frac{(x + \Delta)^2 - x^2}{\Delta} = \frac{x^2 + 2 * x * \Delta + \Delta^2 - x^2}{\Delta}$$

$$= \frac{2 * x * \Delta + \Delta^2}{\Delta} = 2 * x + \Delta$$

We can again take the limit as $\Delta \to 0$ to get:

$$m = 2 * x$$

This result provides the slope of the curve at *any* value of x! In particular, as we already know, $m \mid_{x=2} = 2 * 2 = 4$.

This result for the slope of the curve is in fact the *derivative* of the curve at point x. In fact, for a function $y = f(x)$ of a single variable, x, y can be interpreted as a curve and the *derivative of the function is simply the slope of the curve.* This is conveniently written [4] as

$$derivative = dy/dx = df(x)/dx,$$

suggestive of the fact that the derivative can be thought of as the slope $(\Delta y/\Delta x)$.

Another common shorthand notation for the derivative is

$$derivative = dy/dx = y'.$$

3.3. Linearity of Differentiation

In the Section 3.5 we will find an expression for the derivative of any polynomial of a single variable. First we need to demonstrate an important property of the derivative.

Let's assume we want to take the derivative of a weighted sum of two functions:

$$y = a * f(x) + b * g(x)$$

where a and b are constants. Then simply applying what we know

$$dy/dx =$$

$$lim \mid_{\Delta \to 0} \frac{a * f(x + \Delta) + b * g(x + \Delta) - (a * f(x) + b * g(x))}{\Delta}$$

Rearranging, this can be written as

$$dy/dx = a * [lim \mid_{\Delta \to 0} \frac{f(x + \Delta) - f(x)}{\Delta}] +$$

$$b * [lim \mid_{\Delta \to 0} \frac{g(x + \Delta) - g(x)}{\Delta}]$$

which can be recognized as just

$$dy/dx = a * df/dx + b * dg/dx.$$

Note that the constants a and b are not involved in the limit and therefore can be taken out as shown.

In words, *the derivative of the weighted sum of two functions is the weighted sum of the individual derivatives.* This is characteristic of linear [6] operations, such as the derivative. We will use this property many times.

> Worth Remembering:
> Derivatives are linear, so that
> If
>
> $$y(x) = a * f(x) + b * g(x)$$
>
> then
>
> $$dy(x)/dx = a * df(x)/dx +$$
>
> $$b * dg(x)/dx$$

3.4. Derivative of the Sin(x) and Cos(x) Functions

To find the derivative of the $\sin(x)$ we can just use the definition of a derivative again:

$$\frac{d(\sin(x))}{dx} = lim \mid_{\Delta \to 0} \frac{\sin(x + \Delta) - \sin(x)}{\Delta}.$$

Recall from Section 1.6 on page 14 the following double angle formula

$$\sin(x + \Delta) = \sin(x) * \cos(\Delta) + \cos(x) * \sin(\Delta)$$

which, realizing that $lim \mid_{\Delta \to 0} \cos(\Delta) = 1$ and $lim \mid_{\Delta \to 0} \sin(\Delta) = \Delta$, allows us to write

$$\frac{d(\sin(x))}{dx} = lim \mid_{\Delta \to 0} \frac{\cos(x) * \Delta}{\Delta} = \cos(x).$$

The same approach can be used to quickly show that

$$\frac{d(\cos(x))}{dx} = -\sin(x).$$

Worth Remembering:

$$d(\sin(x))/dx = \cos(x)$$

$$d(\cos(x))/dx = -\sin(x)$$

3.5. Derivative of a Polynomial

Section 3.1 on page 35 outlined a method to calculate the derivative of the function $y = x^2$ with respect to the variable x. In the notation introduced in Section 3.1 on page 35 this derivative is usually written as

$$dy/dx = d(x^2)/dy$$

and has been shown to be $dy/dx = 2 * x$.

This method can be generalized to an arbitrary polynomial in one variable. Consider

$$y = \sum_{i=-\infty}^{i=\infty} a_i * x^i$$

With appropriate selection of the constant coefficients, a_i, this can represent *any* polynomial and many useful functions can be written in terms of an infinite polynomial sum. Now consider taking the derivative of this polynomial. Since we know that the derivative of a sum of functions is the sum of the derivatives, we can examine a single term in the sum, such as

$$dy_i/dx = d(a_i * x^i)/dx = a_i * d(x^i)/dx$$

3. The Derivative

i	$\binom{i}{j}$
0	1
1	1 1
2	1 2 1
3	1 3 3 1
4	1 4 6 4 1
5	1 5 10 10 5 1

Table 3.2.: Binomial Coefficients

$$= a_i * lim \, |_{\Delta \to 0} \frac{[(x + \Delta)^i - x^i]}{\Delta}.$$

Focus on the term $lim \, |_{\Delta \to 0} \frac{[(x+\Delta)^i-x^i]}{\Delta}$. We will show that this is equivalent to $i * x^{i-1}$. Note that

$$(x + \Delta)^i = b_0 * x^i + b_1 * x^{i-1} * \Delta + b_2 * x^{i-2} * \Delta^2 + ...$$

$$... + b_{i-1} * x * \Delta^{i-1} + b_i * \Delta^i$$

where the $b_j \, |_{,j=0,...,i}$ are the binomial coefficients $\binom{i}{j}$ and we have used the Binomial Expansion [7].

Note that $b_0 = 1$ so that in the binomial expansion for $(x + \Delta)^i$, it is clear that in

$$lim \, |_{\Delta \to 0} \frac{[(x + \Delta)^i - x^i]}{\Delta}$$

all terms will vanish in the limit except $b_1 * x^{i-1}$. Table 3.2 lists the first few sets of binomial coefficients. The pattern

Polynomial $[y(x)]$	Derivative $[dy(x)/dx]$
x^2	$2*x$
$3*x^5$	$15*x^4$
$5*x^3 - 8*x^{-4}$	$15*x^2 + 32*x^{-5}$
$\sqrt{3}*x^{-\sqrt{7}}$	$-\sqrt{21}*x^{-\sqrt{7}-1}$

Table 3.3.: Polynomial Derivatives

is obvious. Note that the second term (b_1) in the set i equals i, so that we can write

$$d(x^i)/dx = i*x^{i-1}. \qquad (3.1)$$

What a great result! We can now write down the derivative of any polynomial of a single variable! Table 3.3 shows some examples. It develops that the binomial expansion works for any power, including 0, negative numbers, and even non-integer powers, so this result is very useful!

Worth Remembering:

$$d(x^i)/dx = i*x^{i-1}$$

To return to the issue at hand, the derivative of a polynomial can now be seen to be:

$$dy(x)/dx = d(\sum_{i=-\infty}^{i=\infty} a_i*x^i)/dx = \sum_{i=-\infty}^{i=\infty} a_i*d(x^i)/dx$$

$$= \sum_{i=-\infty}^{i=\infty} a_i*i*x^{i-1}$$

3.6. Deriving Derivatives using Series Expansions

The result in Section 3.5 is really useful. Many common functions such as $\cos(x)$, $\sin(x)$, $\exp(x)$, and many others can be expressed in terms of a polynomial. If we accept the series expansion (polynomial) definitions of a function we can often obtain the derivative of that function easily. Let's get the derivatives of $\sin(x)$, $\cos(x)$, and $\exp(x)$ using Series Expansion.

3.6.1. Derivative of Sin(x) using Series Expansion

Consider [8]

$$\sin(x) = \sum_{n=0}^{n=\infty} \frac{(-1)^n x^{(2n+1)}}{(2n+1)!}.$$

To obtain the derivative of $\sin(x)$ we can just apply the result from Equation 3.1 on the preceding page to each term

$$d(\frac{(-1)^n x^{(2n+1)}}{(2n+1)!})/dx = \frac{(-1)^n (2n+1) * x^{2n}}{(2n+1)!} = \frac{(-1)^n x^{2n}}{(2n)!}$$

so that

$$d(\sin(x))/dx = \sum_{n=0}^{n=\infty} \frac{(-1)^n x^{2n}}{(2n)!}$$

which is exactly the series expression for $\cos(x)$! We therefore have obtained the result from Section 3.4 on page 40 using the series expansion:

$$d(\sin(x))/dx = \cos(x).$$

3.6.2. Derivative of Cos(x) using Series Expansion

In the same way, it is easy to find the derivative of the $\cos(x)$. Consider

$$\cos(x) = \sum_{n=0}^{n=\infty} \frac{(-1)^n x^{2n}}{(2n)!} = 1 + \sum_{n=1}^{n=\infty} \frac{(-1)^n x^{2n}}{(2n)!}.$$

Now, applying Equation 3.1 on page 43,

$$d(\cos(x))/dx = \sum_{n=1}^{n=\infty} \frac{(-1)^n x^{(2n-1)}}{(2n-1)!}$$

(remember that $1 = x^0$ so that when Equation 3.1 on page 43 is used it becomes 0)

which becomes, letting $m = n - 1$,

$$= \sum_{m=0}^{m=\infty} \frac{(-1)^{m+1} x^{2m+1}}{(2m+1)!}$$

$$= (-1) * \sum_{m=0}^{m=\infty} \frac{(-1)^m x^{(2m+1)}}{(2m+1)!}$$

$$= -\sin(x)$$

again repeating the result from Section 3.4 on page 40.

3.6.3. Derivative of the Exponential Function using Series Expansion

The exponential function [9] can be written as the series

$$\exp(x) = e^x = \sum_{n=0}^{n=\infty} \frac{x^n}{n!}.$$

To get the derivative we can apply Equation 3.1 to each term, resulting in

$$d(\exp(x))/dx = \sum_{n=0}^{n=\infty} \frac{n * x^{n-1}}{n!}$$

$$= \sum_{n=1}^{n=\infty} \frac{x^{n-1}}{(n-1)!}$$

(the term for $n = 0$ vanishes, remember that 0!=1) which, replacing $n - 1$ with n in fact is exactly the same as

$$= \sum_{n=0}^{n=\infty} \frac{x^n}{n!} = \exp(x)$$

which is the result we have seen before in Section 3.4 on page 40.

3.7. Derivative of an Exponential

We can obtain the derivative of the exponential function $\exp(x)$ using the definition of the derivative as follows:

$$d(\exp(x))/dx = \lim |_{\Delta \to 0} \frac{\exp(x + \Delta) - \exp(x)}{\Delta}.$$

Recall that $\exp(a + b) = \exp(a) * \exp(b)$ so we can write

$$d(\exp(x))/dx = \lim |_{\Delta \to 0} \frac{\exp(x) * \exp(\Delta) - \exp(x)}{\Delta}$$

$$= \exp(x) * \lim |_{\Delta \to 0} \frac{\exp(\Delta) - 1}{\Delta}$$

$$= a * \exp(x)$$

where the constant a is given as

$$a = \lim |_{\Delta \to 0} \frac{\exp(\Delta) - 1}{\Delta}.$$

This is an interesting result - the derivative of the exponential function is simply proportional to the function itself!

The constant a is determined from a limit process. We need to know how $\exp(\Delta)$ behaves as Δ becomes small. From [9] we have

$$\exp(\Delta) = e^{\Delta} = \sum_{n=0}^{n=\infty} \frac{\Delta^n}{n!}.$$

Clearly as $\Delta \to 0$ the higher order terms vanish quickly. Taking the first two terms (since higher ones involve Δ^2 and smaller terms) gives us

$$\exp(\Delta) \simeq 1 + \Delta$$

so we have for a:

$$a = \lim |_{\Delta \to 0} \frac{1 + \Delta - 1}{\Delta} = 1.$$

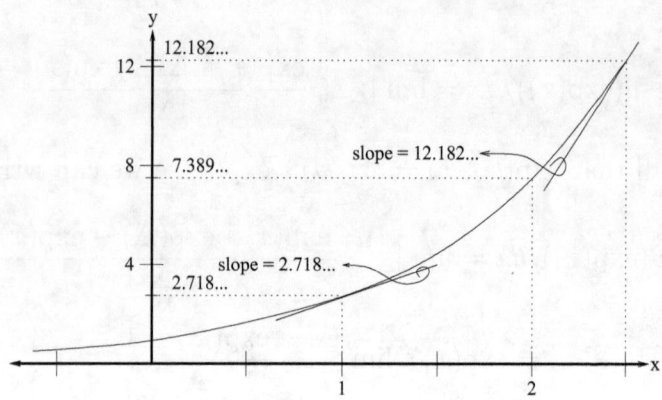

Figure 3.1.: Exponential Function

So, in words, *the derivative of e^x is e^x*, or, in our preferred notation:

$$d(e^x)/dx = e^x.$$

Here's another way to think about the exponential function: when written as $y = e^x$, the value of e^x at any value of x is the slope of the curve y at that value of x. Figure 3.1 shows this unique function.

We can generalize our result for the derivative of $\exp(x)$ by allowing a constant multiplier of the variable:

$$d(\exp(b * x))/dx = \lim |_{\Delta \to 0} \frac{\exp(b * (x + \Delta)) - \exp(b * x)}{\Delta}$$

$$= \exp(b * x) * \lim |_{\Delta \to 0} \frac{\exp(b * \Delta) - 1}{\Delta}.$$

Using the same two term approximation for $\exp(b * \Delta)$ that we used before for $\exp(\Delta)$:

$$d(\exp(b * x))/dx == \exp(b * x) * \lim |_{\Delta \to 0} \frac{1 + (b * \Delta) - 1}{\Delta}$$

$$b * \exp(b * x)$$

This same result can be arrived at using what we know about the derivative when the variable is scaled (Section: 3.21):

$$d(exp(b * x))/dx = b * \frac{d(exp(z))}{dz} \big|_{z=b*x} = b * exp(b * x).$$

Worth Remembering:

$$d(\exp(b*x))/dx = b*\exp(b*x)$$

3.8. Motivational Problem 2.2 - Solution

At $t = 0$ the switch in Figure 2.3 closes and the voltage across the capacitor, initially V_o Volts, is put across the resistor R, causing a current flow through the resistor and out of the capacitor of V_o/R Amps. This current flow will remove charge from the capacitor, reducing the voltage and hence the current. The higher the current the faster the charge will be removed, so the rate of discharge of the capacitor is proportional to the current $I(t)$:

$$dV(t)/dt = -\lambda * I(t).$$

The negative sign reflects that increased current increases the rate of *discharge*. Recall that because of Ohm's Law the current is also proportional to $V(t)$ so that we can write:

$$d[I(t) * R]/dt = -\gamma * I(t)$$

or

$$dI(t)/dt = -\lambda * I(t).$$

So, we are looking for a function of time whose slope is proportional to the value of the function - just like exp(t)! It makes sense to try something like:

$$I(t) = I_o * \exp(a * t) \tag{3.2}$$

How do we find the two constants I_o and a? We know what the initial current is at $t = 0$ - it is simply (V_o/R) so we can write equation 3.2 as

$$I(0) = I_o * \exp(a * 0) = I_o = V_o/R.$$

If we think about it, we also know the time rate of change of the current at $t = 0$, namely $dI(0)/dt$. Remember that $dQ(t)/dt = -I(t)$ so that

$$dQ(0)/dt = -I_o,$$

but since $Q(t) = C * V(t) = C * I(t) * R$ we can write

$$dI(0)/dt = -I_o/(RC).$$

Differentiating equation 3.2 gives

$$dI(t)/dt = a * I_o * \exp(a * t)$$

or

$$dI(0)/dt = a * I_o = -I_o/RC$$

so that $a = -1/RC$ and we can finally write the solution to Motivational Problem 2.2 as:

$$I(t) = (V_o/R) * \exp(-t/RC).$$

Note that the quantity $R * C$ has units of time and is commonly referred to as the *time constant* of the circuit.

3.9. Motivational Problem 2.3 - Solution

Remember Motivational Problem 2.3 (Section 2.3 on page 21)? We were given

$$dN(t)/dt = -\lambda * N(t).$$

Note that N is a function (of time) so that we are looking for a function $N(t)$ whose derivative is proportional to the function itself - just like $\exp(x)$!

Relating this to Motivational Problem 2.3, it seems reasonable that since $N(t)$, the function we need to solve this problem, has

$$dN(t)/dt = -\lambda * N(t)$$

we should try

$$N(t) = N_0 exp(-\lambda * t)$$

where N_0 is the number of radioactive atoms at $t = 0$ (the time the earth was formed). By the way, we just solved a differential equation!

From the problem definition, we have:

$$^{235}N_0 = ^{238}N_0 = N_0.$$

Let T be the age of the earth as determined by the isotope ratio (137.8). We can write

$$\frac{^{238}N(T)}{^{235}N(T)} = 137.8 = \frac{N_o * exp(-\lambda_{238} * T)}{N_o * exp(-\lambda_{235} * T)}$$

$$= exp(T * (-1.55E - 10 + 9.80E - 10))$$

which is easily solved for $T = 5.97$ Billion years.

3.10. Derivative of the Logarithm

Let's look at

$$y(x) = \log_b(a * x)$$

and see if we can find $d(y(x))/dx$. Recall that the notation $y(x) = \log_b(a * x)$ means that $y(x)$ is the power that the base b must be raised to in order to equal $a * x$, in other words $a * x = b^{y(x)}$.

Using the definition of the derivative we have:

$$d(y(x))/dx = \lim |_{h \to 0} \frac{y(x+h) - y(x)}{h}$$

$$= \lim |_{h \to 0} \frac{\log_b(a * (x+h)) - \log_b(a * x)}{h}$$

$$= \lim |_{h \to 0} \frac{\log_b(a) + \log_b(x+h) - \log_b(a) - \log_b(x)}{h}$$

$$= \lim |_{h \to 0} \frac{1}{h} * \log_b(\frac{x+h}{x})$$

$$= \lim |_{h \to 0} \log_b[(1 + \frac{h}{x})^{1/h}]$$

Substituting $h/x = s$, we write

$$= \lim |_{s \to 0} \log_b[(1 + s)^{1/(s*x)}]$$

$$= \lim |_{s \to 0} \log_b\{[(1 + s)^{1/s}]^{1/x}\}$$

or, since $\lim |_{s \to 0} (1 + s)^{1/s} = e$ (see Section 1.4 on page 9),

$$d(\log_b(a * x))/dx = \log_b(e^{1/x})$$

$$= (1/x) * \log_b(e).$$

It may seem strange that the derivative does not depend on the constant a, but consider that

$$\log_b(a * x) = \log_b(a) + \log_b(x)$$

and since the derivative of a constant is zero the result is expected.

If we use e, the base of the natural logarithms, this simplifies to

$$d(\ln(a * x))/dx = \ln(e^{1/x})$$

$$= (1/x) * \ln(e)$$

$$= 1/x.$$

> **Worth Remembering**
>
> $$d(\log_b(a * x))/dx = (1/x) * \log_b(e)$$
>
> $$d(\ln(a * x))/dx = 1/x$$

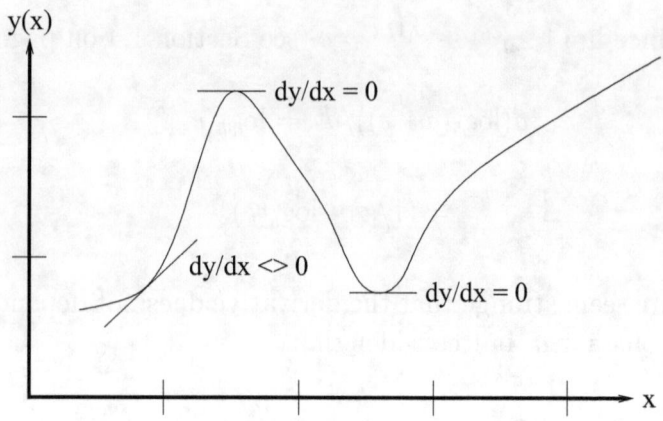

Figure 3.2.: Minimum and Maximum of a Curve

3.11. Minimum and Maximum Values

Recognizing the derivative as the slope of a curve suggests a way to find the minimum or maximum of a function. Figure 3.2 illustrates the basic idea. Given a curve $y(x)$ it is clear that the values of x that make the derivative (slope) go to zero are the very values of x at which the curve has at least a local minimum or maximum.

Note that some curves will throw you a curve where a zero slope will be an inflection point and not a true minimum of maximum. A simple example is the function $y(x) = x^3$. The derivative is $dy/dx = 3 * x^2$, which goes to zero at $x = 0$. In spite of this result, the plot of $y(x)$ (see Figure 3.3) clearly shows that there is no minimum or maximum at $x = 0$.

It is always a good idea to check your results for consistency and reasonableness!

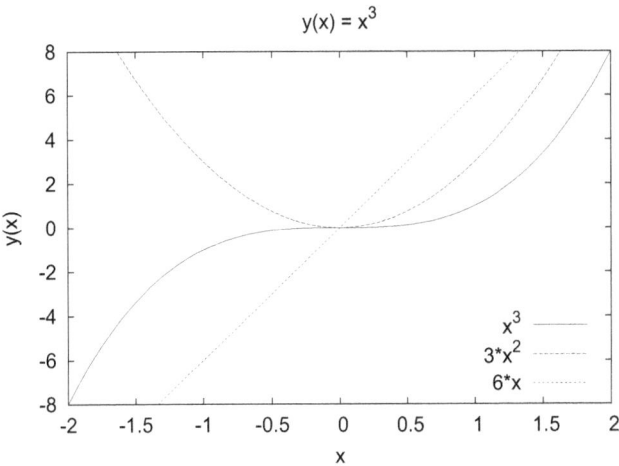

Figure 3.3.: $y(x) = x^3, y'/dx = 3 * x^2, y'' = 6 * x$

The thing about inflection points is that the slope, although zero at the critical point, will have the same sign on both sides of the point.

True minimums and maximums will have slopes with different signs on the two sides of the point with zero slope. As an example, the derivative of $3 * x^2$ is $6 * x$ and is plotted in Figure 3.3, showing that the function $3 * x^2$ has a minimum at $x = 0$.

Table 3.4 summarizes all this for an assumed critical point at $x = a$.

Type	$y'(x < a)$	$y'(x = a)$	$y'(x > a)$
Min	−	0	+
Max	+	0	−
Inflection+	+	0	+
Inflection-	−	0	−

Table 3.4.: Min/Max and Inflection Points

3.12. Motivational Problem 2.4 - Solution

With this concept in mind, we are ready to solve Motivational Problem 2.4. The final design is to be a rectangle, so 2 sides will be the same length, say L ft., and the other two sides will be the same length, say, M ft. We can therefore write an expression for the perimeter P of the pasture and its area A:

$$P = 2 * L + 2 * M$$

$$A = L * M.$$

We want to maximize the area A subject to the constraint that the total cost of fencing is not to exceed the budgeted amount $\$B$. Fencing costs $\$F/ft$. Using this information we can write the perimeter as

$$P = B/F$$

then

$$M = (P/2) - L = (B/2F) - L$$

so that

$$A(L) = L * (K - L) = K * L - L^2$$

where $K = B/(2F)$, a constant. We can write all this as a function y of x:

$$y(x) = K * x - x^2$$

Recall that the derivative of a sum of functions is the sum of the derivatives, and that the derivative is just the slope of a curve. Combining this with the result from Section 2.1, we can write

$$\frac{d(y(x))}{dx} = K - 2 * x$$

which we set to zero to find the value of x that maximizes the pasture:

$$0 = K - 2 * L$$

so that the optimum value of L is

$$L = K/2 = B/(4F)$$

resulting in a square since

$$M = (B/2F) - L$$

$$= B/(4F)$$

$$= L.$$

The area of this square pasture can be expressed as:

$$A = L * M = L^2 = P^2/16$$

recalling that the perimeter $P = B/F$.

3.13. An Even Better Pasture?

A rectangular pasture may not be optimum if the cost of the fencing is the only criteria used. Consider a circular pasture - we know that the circumference of a circle is $2 * \pi * R$ and the area is $\pi * R^2$ where R is the radius of the circle. Relating this to Motivational Problem 2.4 we would set the perimeter equal to the circle's circumference:

$$P = B/F = 2 * \pi * R$$

from which we can express the area as:

$$A = P^2/(4 * \pi)$$

Comparing this to the result in Section 3.12 we see an improvement in area of $4/\pi$ or about 27%. So, why aren't more pastures circular rather than rectangular?

3.14. Motivational Problem 2.5 - Solution

The profit we make from the sale of Widgets will be the price we obtain times the number of units we can sell minus all of our costs. In terms of the problem definition (see Section 2.5 on page 22):

$$Profit = W * (\$10.00 - \$0.025 * W) - (\$500.00 + \$2.5 * W)$$

$$= \$10.00 * W - \$0.025 * W^2 - \$500.00 - \$2.50 * W$$

To maximize the profit we set the derivative of the *Profit* with respect to W equal to zero and solve for W :

$$d(Profit)/dW = \$10.00 - \$0.05 * W - \$2.50 = 0$$

or

$$W = \$7.50/\$0.05 = 150.$$

We can approach this from another direction[13]. At any given number of Widgets there is a marginal cost - the cost to produce the next Widget. There will also be a marginal revenue - the revenue we obtain from selling it. At any number of Widgets, if the marginal revenue is greater than the marginal cost, we should make more Widgets. If the marginal cost is greater than the marginal revenue, we should make less Widgets. Clearly our profit will be maximized when the marginal cost is equal to the marginal revenue. Of course, the derivative of the cost function is the marginal cost and the derivative of the revenue function is the marginal revenue. In terms of our example:

$$Cost(W) = \$500.00 + \$2.50 * W$$

$$Revenue(W) = W * (\$10.00 - \$0.025 * W)$$

$$dCost(W)/dw = \$2.50$$

$$dRevenue(W)/dw = \$10.00 - \$0.05 * W$$

Setting the marginal cost equal to the marginal revenue gives us the optimum number of Widgets to produce - 150 as we expected.

According to the demand curve we can sell the 150 Widgets if we offer them at $6.25 each for a total income of $937.50.

It will cost us $875.00 to make 150 Widgets so our total profit will be $62.50 or a little over a 7% return on our $875.00 investment. Maybe we should consider something else to make!

Note that the fixed cost of $500.00 doesn't enter into the calculation for the optimum number of Widgets, but of course it directly reduces our profit!

3.15. The Product Rule - Derivative of the Product of two Functions

We often are faced with a function that is written as the product of two functions, such as

$$y(x) = exp(-x) * \sin(x).$$

This section will derive the Product Rule that can be used to handle this type of function. Simply applying the definition of the derivative as a limit we proceed as follows:

$$y(x) = f(x) * g(x)$$

where $f(x)$ and $g(x)$ are each a function of x. Then

$$d(y(x))/dx = \lim |_{\Delta \to 0} \frac{f(x + \Delta) * g(x + \Delta) - f(x) * g(x)}{\Delta}.$$

Now, following [12], we add and subtract $f(x + \Delta) * g(x)$ to the numerator

$$= \lim |_{\Delta \to 0} \{ \frac{f(x + \Delta) * g(x + \Delta) - f(x) * g(x)}{\Delta}$$

$$+\frac{f(x+\Delta)*g(x)-f(x+\Delta)*g(x)}{\Delta}\}$$

(note the {...} - the limit applies to both additive terms)
and rearrange

$$= \lim \Big|_{\Delta\to 0} \{f(x+\Delta)*\frac{(g(x+\Delta)-g(x))}{\Delta}+$$

$$g(x)*\frac{(f(x+\Delta)-f(x))}{\Delta}\}$$

$$= f(x)*\lim \Big|_{\Delta\to 0} \frac{(g(x+\Delta)-g(x))}{\Delta}+$$

$$g(x)*\lim \Big|_{\Delta\to 0} \frac{(f(x+\Delta)-f(x))}{\Delta}$$

$$= f(x)*\frac{d(g(x))}{dx}+g(x)*\frac{d(f(x))}{dx}.$$

In words, *the derivative of the product of two functions is equal to the first function times the derivative of the second plus the second function times the derivative of the first.*

Worth Remembering:
The Product Rule

$$\frac{d(f(x)*g(x))}{dx}=$$

$$f(x)*\frac{d(g(x))}{dx}+g(x)*\frac{d(f(x))}{dx}$$

3.16. Motivational Problem 2.6 - Solution

We can now demonstrate that Euler's Formula is valid [15]. Consider the function

$$f(x) = (cos(x) + j * sin(x)) * e^{-j*x}.$$

First, take the derivative of $f(x)$, using the product rule:

$$d(f(x))/dx = (cos(x) + j * sin(x)) * (-j * e^{-j*x}) +$$

$$e^{-j*x} * (-\sin(x) + j * \cos(x))$$

$$= \exp(-j * x) * [-j * \cos(x) + \sin(x) - \sin(x) + j * \cos(x)]$$

$$= 0.$$

So, the slope of $f(x) = 0$ for all values of x, which means that $f(x)$ must be a constant! We know the value of $f(x)$ for $x = 0$, it is just

$$f(0) = 1 = f(x)$$

so we have

$$1 = (\cos(x) + j * \sin(x)) * e^{-jx}$$

or

$$e^{j*x} = \cos(x) + j * \sin(x)$$

which is Euler's Identity.

3.17. Euler's Identity

We have solved Motivational Problem 2.6 but let's look at this remarkable formula a little more. If we set $x = \pi$ we get

$$e^{j*\pi} = \cos(\pi) + j * \sin(\pi)$$

which can be written as

$$e^{j*\pi} + 1 = 0.$$

Known as Euler's Identity, this expression combines the numbers e, $\sqrt{-1}$, π, 1, and 0 into what must be the most beautiful mathematical expression known!

3.18. The Quotient Rule - Derivative of the Quotient of two Functions

Consider the derivative of the quotient of two functions:

$$d(u/v)/dx$$

where u and v are each functions of x. Simply applying the definition of a derivative gives us

$$d(u(x)/v(x))/dx = \lim |_{h \to 0} \frac{u(x+h)/v(x+h) - u(x)/v(x)}{h}$$

$$= \lim |_{h \to 0} \frac{v(x) * u(x+h) - u(x) * v(x+h)}{h * v(x) * v(x+h)}. \qquad (3.3)$$

3. The Derivative

Consider the meaning of $u(x+h)$ and $v(x+h)$. These are the values of the functions evaluated at $x+h$ so they are equivalent, when h is small, to

$$u(x+h) = u(x) + \frac{du(x)}{dx} * h$$

and

$$v(x+h) = v(x) + \frac{dv(x)}{dx} * h$$

Substituting these into equation 3.3 results in

$$\frac{d(u/v)}{dx} =$$

$$\lim|_{h \to 0} \frac{v(x) * (u(x) + \frac{du}{dx} * h) - u(x) * (v(x) + \frac{dv}{dx} * h)}{h * v(x) * (v(x) + \frac{dv}{dx} * h)}.$$

Note that the terms $v(x) * u(x)$ and $-u(x) * v(x)$ in the numerator cancel leaving

$$= \lim|_{h \to 0} \frac{v(x) * \frac{du}{dx} * h) - u(x) * \frac{dv}{dx} * h}{h * v(x) * v(x) + h * h * \frac{dv}{dx}}$$

$$= (v\frac{du}{dx} - u\frac{dv}{dx})/v^2$$

> **Worth Remembering**
> **The Quotient Rule**
> Given functions $u(x)$ and $v(x)$ we have
> $$\frac{d(u/v)}{dx} = \frac{v * du/dx - u * dv/dx}{v^2}$$

As an example, we can get the derivative of the reciprocal function $1/x$. Let $u = 1$ and $v = x$, then

$$\frac{d(1/x)}{dx} = \frac{x * 0 - 1 * 1}{x^2} = -1/x^2$$

Worth Remembering

$$\frac{d(1/x)}{dx} = -1/x^2$$

3.19. Motivational Problem 2.7 - Solution

Motivational Problem 2.7 involves finding a maximum volume of a package being shipped via UPS. We are given the following information:

$$L \leq 108$$

$$L + 2 * W + 2 * H = 165$$

$$L \geq W$$
$$L \geq H$$

where L is the length (greatest dimension), W is the width, and H is the height. We want to maximize the volume:

$$V = L * W * H.$$

3. The Derivative

We need to express the volume of the package in a way that we can take its derivative so we can set it to zero.

Let $H = a * W$. Then

$$L = 165 - 2 * (1 + a) * W$$

from which

$$V(W) = L * a * W^2 = [165 - 2 * (1 + a) * W] * a * W^2.$$

To find a maximum we take the derivative (with respect to the only remaining variable W) and set it to zero:

$$d(V(W))/dW = a * 330 * W - 6 * a * W^2 - 6 * a^2 * W^2 = 0$$

or

$$330 - 6 * W * (1 + a) = 0.$$

The value of W that maximizes the package volume is therefore

$$W_{max} = 55/(1 + a) \tag{3.4}$$

and the resulting maximum volume is

$$V_{max} = [165 - 2 * (1 + a) * 55/(1 + a)] * a * [55/(1 + a)]^2$$

$$= C * \frac{a}{(1 + a)^2} \tag{3.5}$$

where C has a value of $166,375 in^3$.

Now we have to find the value of a that maximizes the expression $a/(1 + a)^2$.

Let

$$y(a) = a/(1 + a)^2.$$

Again, we need to take the derivative and set it to zero. We can use the Quotient Rule (Section: 3.18) to take the derivative:

$$d(a/(1+a)^2)/da = \frac{(1+a)^2 - a*2*(1+a)}{(1+a)^4} = 0$$

or

$$(1+a) - a*2 = 0$$

so that the a that maximizes the volume is $a = 1$. Using this result in Equation 3.5 finally gives

$$V_{max} = C/4 = 41,593.75in^3 \simeq 24.1ft^3.$$

Recalling Equation 3.4 we have the final package dimensions of $W_{max} = H_{max} = 27.5in$ and $L = 55in$.

3.20. The Chain Rule - Derivative of Composite of two Functions

A composite function is obtained by applying a second function to the result of a first function. If $f(x)$ and $g(x)$ are functions, then we can define the composite g of f as the result of applying function g to the function f. This is usually written as

$$g \circ f = g(f(x)).$$

How can we take the derivative of a composite function? If we change x by a small amount, say Δ_x, then we know that $f(x)$will change by a small amount, say Δ_f. This change in $f(x)$ will in turn cause a change in the function g. Well, how much does $f(x)$ change as a result of x changing by Δ_x? Recalling the definition of a derivative as a slope, it is clear that

$$\Delta_f \simeq \frac{d(f(x))}{dx} * \Delta_x$$

and in fact, in the limit as Δ_x, and hence Δ_f, tend to 0, this becomes exact.

Well, since the function g is a function of $f(x)$, we can similarly estimate the change in g caused by a small change in $f(x)$ as:

$$\Delta_g \simeq \frac{d(g(f))}{df} * \Delta_f$$

or

$$\Delta_g \simeq \frac{d(g(f))}{df} * \frac{d(f(x))}{dx} * \Delta_x. \qquad (3.6)$$

Again, this becomes exact as the Δ's tend to 0.

Of course, what we started out to find was the derivative of $g(f(x))$ with respect to x, not with respect to f, but that is just

$$\lim |_{\Delta_x \to 0} \frac{\Delta_g}{\Delta_x}$$

which, from equation 3.6, is

$$\frac{d(g(f(x)))}{dx} = \frac{d(g(f))}{df} * \frac{d(f(x))}{dx}.$$

In words, *the derivative of a composite function can be found as the product of the derivative of the two functions taken individually.*

> Worth Remembering:
> The Chain Rule
>
> $$\frac{d(g(f(x)))}{dx} = \frac{d(g(f))}{df} * \frac{d(f(x))}{dx}$$

As an example of using the Chain Rule, consider

$$d(ln(x+1))/dx.$$

We have $f(x) = (x+1)$ and $g(f(x)) = ln(f(x))$. Let $y = x + 1$, then

$$d(ln(x+1))/dx = [d(ln(y))/dy] * d(x+1)/dx = 1/y$$

$$= 1/(x+1)$$

This result is easily generalized to

$$d(ln(a * x + b))/x = a/(a * x + b).$$

3.21. Derivative of a Function with Scaled Independent Variable

What happens when we stretch or shrink the independent variable of a function? Consider finding

$$\frac{d(y(a * x))}{dx}$$

where the x variable has been scaled by the constant a.

Applying the chain rule (Section: 3.20) by substituting y for g and $a * x$ for $f(x)$ gives:

$$\frac{d(y(a * x))}{dx} = \frac{d(y(a * x))}{d(a * x)} * \frac{d(a * x)}{dx}$$

$$= a * \frac{d(y(a * x))}{d(a * x)}.$$

What does $d(y(a * x))/d(a * x)$ mean? Note that $a * x$ does not refer to a specific value of a or x, in fact $a * x$ is simply a placeholder that could be just as well written as z. With this in mind, we have

> **Worth Remembering:**
> Derivative with scaled variable
> $$\frac{d(y(a * x))}{dx} = a * \frac{d(y(z))}{dz} \Big|_{z=a*x}$$

In words, the derivative of a function with a scaled variable is a times the derivative of the function with an unscaled variable, evaluated at $a * x$.

As an example, consider

$$\frac{d(\sin(a * x))}{dx} = a * \cos(a * x).$$

Here's another example, one whose result we already have seen:

$$d(\log_b(a * x))/dx = a * d(\log_b(z))/dz \big|_{z=a*x}$$

$$= a * \log_b(e^{1/(a*x)}) = \frac{a}{a} * \frac{1}{x} \log_b(e)$$

which is not a function of the constant a, as we found in Section 3.10 on page 52.

3.22. L'Hôpital's Rule for Indeterminate Limit of Type 0/0

We sometimes need to find limits like

$$\lim|_{x \to a} \frac{f(x)}{g(x)} \qquad (3.7)$$

where $\lim|_{x \to a} f(x) = 0$ and $\lim|_{x \to a} g(x) = 0$. This is known as an indeterminate limit of the type 0/0. Since simply taking the limit of the fraction gives 0/0 how do we evaluate a limit like 3.7?

As a side note, we have been using functions that give the 0/0 form for some time now - consider the definition of a derivative:

$$dy(x)/dx = \lim|_{h \to 0} \frac{y(x + h) - y(x)}{h}$$

which becomes 0/0 if we simply substitute $h = 0$!

Back to the problem at hand - consider the first order approximation to an arbitrary function, say $y(x)$, in the vicinity of $x = a$:

$$y(x) \simeq y(a) + (X - a) * d(y(x))/dx|_{x=a} . \qquad (3.8)$$

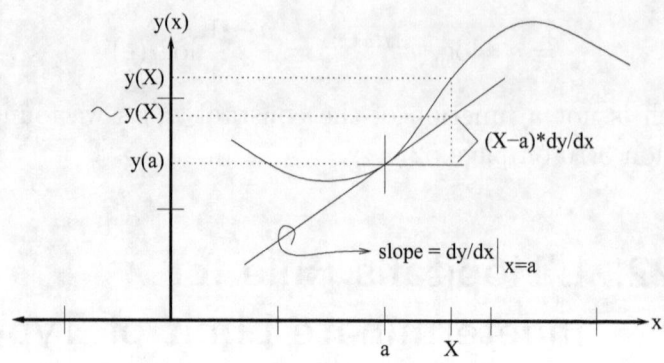

Figure 3.4.: Approximating y(x) near x=a

In words, Equation 3.8 states that we can approximate $y(x)$ for small excursions of x away from $x = a$ by adding the correction term $(x-a)*d(y(x))/dx \,|_{x=a}$ to $y(a)$. This makes a lot of sense when we realize that the derivative is the slope of the curve. Figure (3.4) illustrates the point. Clearly the closer X is to a the better the approximation is.

Using Equation 3.8 for both functions in Equation 3.7 we can write

$$\lim |_{x \to a} \frac{f(x)}{g(x)}$$

$$\simeq \lim |_{x \to a} \frac{f(a) + (x - a) * d(f(x))/dx \,|_{x=a}}{g(a) + (x - a) * d(g(x))/dx \,|_{x=a}}$$

$$= \lim |_{x=a} \frac{(x - a)}{(x - a)} * \frac{d(f(x))/dx}{d(g(x))/dx}$$

$$= \lim |_{x=a} \frac{d(f(x))/dx}{d(g(x))/dx}$$

(recall that $f(a) = g(a) = 0$)

To summarize, if $f(a) = g(a) = 0$ and the right hand limit exists, then

$$\lim \big|_{x=a} \frac{f(x)}{g(x)} = \lim \big|_{x=a} \frac{d(f(x))/dx}{d(g(x))/dx}.$$

This is known as L'Hopital's Rule.

3.23. Motivational Problem 2.8 - Solution

We are asked to find

$$lim \big|_{x \to 0} 3 * \frac{\sin(x)}{x}.$$

Since $f(x) = 3 * \sin(x)/x$ goes to $\frac{0}{0}$ when x goes to 0 we can try L'Hopital's Rule to evaluate this limit.

$$\lim \big|_{x \to 0} 3 * \frac{\sin(x)}{x}$$

$$= 3 * \lim \big|_{x \to 0} \frac{d(\sin(x))/dx}{dx/dx}$$

$$= 3 * \cos(0)/1 = 3.$$

as we suspected from the numerical exercise in Section 1.3 on page 7.

3.24. L'Hôpital's Rule for Indeterminate Limit of the Type ∞/∞

We sometimes need to take a limit of a ration of functions such as

$$L = \lim \big|_{x \to a} \frac{f(x)}{g(x)}$$

where the limit of both $f(x)$ and $g(x)$ is ∞ when $x \to a$. Note that

$$L = \lim \big|_{x \to a} \frac{1/g(x)}{1/f(x)}$$

which is an indeterminate limit of the type $0/0$ which we have already considered. Using the result of Section 3.22 we can immediately write

$$L = \lim \big|_{x \to a} \frac{d(g^{-1}(x))/dx}{d(f^{-1}(x))/dx}$$

$$= \lim \big|_{x \to a} \frac{-d(g(x))/dx}{-d(f(x))/dx} * \frac{f^2(x)}{g^2(x)}$$

$$\lim \big|_{x \to a} \frac{f(x)}{g(x)}.$$

Cross multiplying gives us the desired result:

$$\lim \big|_{x \to a} \frac{f(x)}{g(x)} = \lim \big|_{x \to a} \frac{d(f(x))/dx}{d(g(x))/dx}.$$

To summarize, if $f(a) = g(a) = \infty$ and the right hand limit exists, then

$$\lim \big|_{x=a} \frac{f(x)}{g(x)} = \lim \big|_{x=a} \frac{d(f(x))/dx}{d(g(x))/dx}.$$

3.25. Motivational Problem 2.9 - Solution

We earlier encountered an indeterminate limit in the form of

$$lim\,|_{h\to 0}\,(1+h)^{1/h}.$$

Using the limit value of h, namely 0, results in 1^∞ which is another flavor of indeterminate limit. This limit can be rearranged by taking the ln function of the limit:

$$\ln[\lim\,|_{h\to 0}\,(1+h)^{1/h}] = \lim\,|_{h\to 0}\,\ln[(1+h)^{1/h}]$$

$$= \lim\,|_{h\to 0}\,\ln(1+h)/h.$$

The last term goes to $0/0$ as the limit is taken, so we can apply L'Hôpital's Rule:

$$\lim\,|_{h\to 0}\,\ln(1+h)/h = \lim\,|_{h\to 0}\,[d(\ln(1+h))/dh]/(dh/dh)$$

$$= \lim\,|_{h\to 0}\,1/(1+h) = 1.$$

Reviewing all of this, we can see that

$$\ln[\lim\,|_{h\to 0}\,(1+h)^{1/h}] = 1.$$

By definition, the number whose natural logarithm is 1 is the number e.

Part III.

Integral Calculus

4. The Integral

4.1. The Integral - the Area under a Curve

Just as differentiation is all about slopes and differences, integration is all about areas and sums.

Integration is often used to find the area under a curve. For example, in Motivational Problem 2.12 if we find the area under the curve shown in Figure 2.7 we will know the distance traveled by the car. So, the question is how do we find the area?

4.1.1. Area Using the Riemann Sum

One approach uses what is called the Riemann Sum - see [16] for a web based demonstration and [17, p 1559] for a more complete description. Figure 4.1 shows a curve with the t axis divided into 8 segments, each of which contains a rectangle that approximates the area under the curve for that segment. In general the rectangles do not have to be the same width and the height can be any value that the function $y(t)$ takes within the segment. For convenience the Figure shows equal width segments with bases $\Delta t = (b-a)/8$ and the height of each segment is the value of $y(t)$ at the midpoint of the corresponding segment.

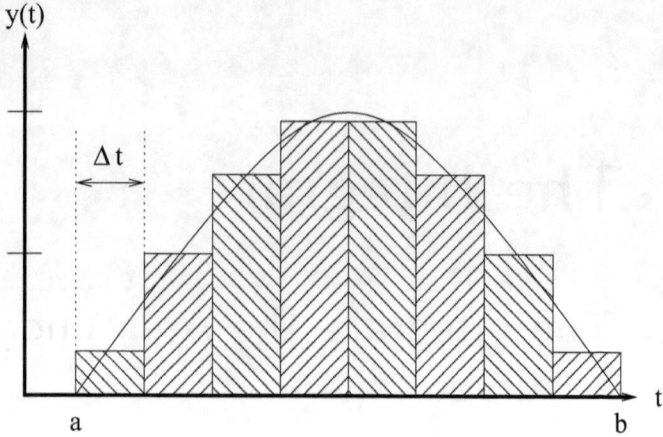

Figure 4.1.: Dividing the Time axis into Segments

If we add up the areas in the 8 rectangles we will have an approximate value for the area under the curve. Although tedious, this can be done. Clearly the approximation could be improved by using more segments in the approximation. If we select N evenly spaced segments as indicated in Figure 4.1 we can write

$$Area \simeq \Delta t * \sum_{i=0}^{i=N-1} y(a + \Delta t/2 + i * \Delta t)$$

and if we let the number of segments grow without bound our approximation becomes exact - that is

$$Area = \lim |_{N \to \infty} \Delta t * \sum_{i=0}^{i=N-1} y(a + \Delta t/2 + i * \Delta t)$$

This expression is the integral of $y(t)$ with respect to t between a and b. This can be written in a more compact notation as

$$Area = \int_a^b y(t)dt$$

which is read as "the integral of $y(t)$ from $t = a$ to $t = b$. Note that the resulting area in *not* a function of t.

4.1.2. A Better Way - The Fundamental Theorem of Calculus

Surely there must be a better way to calculate an integral (the area under a curve) than using a Riemann Sum! There is, and the Fundamental Theorem of Calculus (FToC) shows us how. The FToC comes in two parts, so let's look at the First Fundamental Theorem of Calculus (FFToC) first.

The FFToC states a very useful relationship between the integral of a function and the function itself. Let's define the area function $F(x)$ as

$$F(x) = \int_a^x f(t)dt.$$

This is simply the integral, or the area under the curve $f(t)$, between the (arbitrary, fixed) point a and the variable point x. Note that this will be a function of x as our notation indicates. The variable t is the variable of integration and does not appear in the result of the integration. See Figure 4.2 and note that if the range (from a to x) over which we calculate the area under $f(t)$ is changed by changing x to $x + h$ it is clear that the area, and hence $F(x)$, changes. On the other hand, $F(x)$ is not a function of the dummy variable t as this only serves to indicate that we are taking

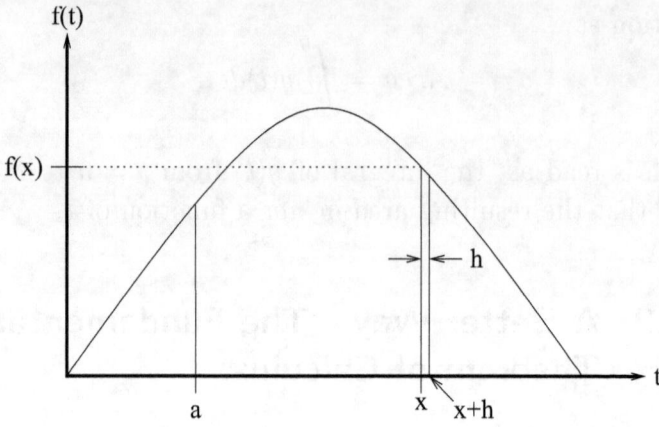

Figure 4.2.: Area Under the Curve from a to x.

the area under the curve. It is true that $F(x)$ depends on the constant a, but we don't specifically call attention to this fact as we are taking a as a constant.

Now, here is the important result from this line of reasoning. The *rate* at which the area changes as x changes is given directly by the height of the curve at x, that is $f(x)$. To see this, consider the area of the thin sliver in Figure 4.2. Recall that the area function, $F(x)$, is the area under the curve between $t = a$ and $t = x$, therefore the area of our sliver, approximately $f(x) * h$, is the difference between $F(x + h)$ and $F(x)$, that is:

$$F(x + h) - F(x) \simeq f(x) * h.$$

If we take the limit as $h \to 0$ this becomes exact and we can write

$$\lim |_{h \to 0} \frac{F(x + h) - F(x)}{h} = f(x).$$

Wait a minute, we have seen this before! From our study of derivatives, the left hand side is nothing more than the derivative of $F(x)$, and our expression can be written as

$$\frac{d(F(x))}{dx} = f(x).$$

This is the FFToC. Note that while $F(x)$ depends on the choice of a, the derivative $d(F(x))/dx$ does not. In words the FFToC states that the derivative of an area function $F(x)$ evaluated at x is the value of the function being integrated evaluated at x.

The area function $F(x)$ defined for the integrand $f(x)$ is called the anti-derivative of $f(x)$.

There is a second part to the FToC, the Second Fundamental Theorem of Calculus (SFToC) that will be useful as well. Consider Figure 4.1 again. Lets add some more notation by labeling the t locations of the rectangular bases:

$$t_i = a + i * \Delta t$$

With the bases specified, the area under the curve, given an approximation using N rectangles and using the value of $y(t)$ at the middle of each rectangle, is:

$$Area \simeq \sum_{i=0}^{i=N-1} y(t_i + \Delta t/2) * \Delta t$$

We can recognize this as a sum of the area of small rectangles. Now, if we know the area function for the given $y(t)$, that is if we know an $F(t)$ such that

$$d(F(t))/dt = y(t)$$

then we can rewrite the sum by noting that the area of each small rectangle can be expressed as

$$y(t_i + \Delta t/2) * \Delta t = F(t_{i+1}) - F(t_i).$$

The sum then becomes

$$Area \simeq \sum_{i=0}^{i=N-1} F(t_{i+1}) - F(t_i).$$

In the limit as $\Delta t \to 0$ this becomes exactly the area we are looking for. Note that the sum can now be seen to be:

$$\sum_{i=0}^{i=N-1} F(t_{i+1}) - F(t_i) = [F(t_1) - F(t_0)] + [F(t_2) - F(t_1)] + ...$$

$$... + [F(t_{N-1}) - F(t_{N-2})] + [F(t_N) - F(t_{N-1})]$$

which conveniently collapses to

$$= F(t_N) - F(t_0) = F(b) - F(a)$$

Finally, taking the limit as $\Delta t \to 0$ and recognizing that the sum becomes an integral, the SFToC states:

$$\int_a^b y(t)dt = F(b) - F(a).$$

This is a great result! If we can find the anti-derivative function $F(x)$ for the integrand $y(x)$ it is trivial to evaluate a definite integral (an integral with specified limits of integration $t = a$ and $t = b$).

Worth Remembering
The First Fundamental Theorem of Calculus:
If we have

$$F(x) = \int_a^x f(t)dt$$

then

$$\frac{d(F(x))}{dx} = f(x)$$

Second Fundamental Theorem of Calculus:
For $F(x)$ defined as above

$$\int_a^b y(t)dt = F(b) - F(a)$$

4.2. Two Types of Integrals

To get a numerical answer to a problem we use the so-called definite integral. This is just an integral that is evaluated between two limits as shown in the SFToC and can be interpreted as the area under the curve between the two limits.

There is also an indefinite integral - simply an integral with no specified limits.

4.2.1. Definite Integral

Definite integrals are evaluated using the SFToC:

$$\int_a^b y(t)dt = F(b) - F(a).$$

The number that results can be interpreted as the area under the curve $y(t)$ between the two limits $x = a$ and $x = b$.

In order to evaluate a definite integral we need to be able to find the anti-derivative of the function that is being integrated. In the notation of the FFToC,

$$\frac{d(F(x))}{dx} = f(x),$$

$F(x)$ is the anti-derivative of $f(x)$. We already know some anti-derivatives from working with derivatives. For example, we know that

$$d(x)/dx = 1$$

so that x must be an anti-derivative of 1. We can immediately evaluate

$$\int_4^2 1 * dx = x \mid_{x=2} - x \mid_{x=4} = -2.$$

Note that $x + c$, where c is a constant, is also an anti-derivative of 1 since $d(x + c)/dx = 1$. There are actually an infinite number of anti-derivatives that differ only by a constant, but note that as far as the SFToC is concerned the constant cancels out.

As another example, we know that

$$d(a * \sin(b * x))/dx = a * b * \cos(b * x)$$

so that $a * \sin(b * x)$ must be an anti-derivative of $a * b * \cos(b * x)$. Of course, so is $a * \sin(b * x) + c$. Appendix A lists several derivative - anti-derivative pairs.

4.2.2. Indefinite Integral

An indefinite integral does not yield a numerical answer as there are no specified limits available to evaluate the integral. Consider

$$\int 1 * dx.$$

How can we interpret such an expression? We know that x is an anti-derivative of 1, so is it reasonable to write:

$$\int 1 * dx = x?$$

Well, almost. As noted above (Section 4.2.1) x is only one of infinitely many anti-derivatives of 1.

In fact, for any given $f(x)$ there will be an infinite number of anti-derivatives, differing only by a constant. This entire group of anti-derivatives can be conveniently represented as an indefinite integral. That is, the indefinite integral $\int 1*dx$ represents the set of all anti-derivatives $x + a$ where a is a constant.

Note that indefinite integrals yield functions while definite integrals yield numerical results.

4.3. Motivational Problem 2.10 - Solution

Let's go ahead and find the equations of motion for an object subject to a constant acceleration a. Newton tells us that force and acceleration are proportional, that is

$$F = M * a$$

where F is the force applied to an object of mass M which then experiences an acceleration of a. If F and M are constant, then so is a, in fact $a = F/M$. With constant acceleration an object will pick up speed at a constant rate. Starting from a standstill, after Δt seconds our object would have a speed of $v = a * \Delta t = \frac{F * \Delta t}{M}$. Lets consider the object under acceleration for a total time of T seconds, split into N intervals of T/N seconds. After N such time intervals (T seconds) the speed would increase to:

$$v = \frac{F}{M} * \sum_{i=0}^{i=N-1} \Delta t_i \qquad (4.1)$$

where the time interval, Δt_i, equals T/N and T is the total time that the object experiences the acceleration. This looks like a Riemann Sum which we can interpret as the area under a curve. Taking a limit where $\Delta t \to 0$ and noting that as i runs from 0 to N that t runs from 0 to T we can write:

$$v = \frac{F}{M} * \int_0^T 1 * dt$$

We need the anti-derivative of 1 (with respect to t) which is simply t (see Section 4.2.1) so we have

$$v = \frac{F}{M} * [t] \,|_{t=0}^{t=T}$$

or, as expected (!), simply

$$v = \frac{F}{M} * T.$$

This seems like a lot of trouble to prove such a simple idea, but the steps we followed can also be used to derive a much more general and interesting result. What happens if the acceleration is not constant? We can indicate this by letting the acceleration, a, be a function of time, say $a(t)$. Following the above steps we arrive at

$$v = \sum_{i=0}^{i=N-1} a(t_i) * \Delta t_i$$

Where we have made the implied assumption that the acceleration during the time interval Δt_i is at least approximately constant at the value of $a(t_i)$. This approximation clearly improves as $\Delta t_i \to 0$. Taking this limit, and noting as before how t behaves as i runs from 0 to $N - 1$, our expression for v under a variable acceleration $a(t)$ becomes:

$$v = \int_0^T a(t) * dt.$$

As long as $a(t)$ has an anti-derivative it is straightforward to get the speed of an object after T seconds even when subjected to a variable acceleration.

With just a little thought we can further generalize this result. Let's say we know the speed at some initial time, say T_1, and the acceleration is applied from T_1 to some final time T_2. Clearly we can get the speed at T_2 by starting with $v(T_1)$ and adding the effect of the acceleration:

$$v(T_2) = v(T_1) + \int_{T_1}^{T_2} a(t) * dt.$$

Well, given a speed, how do we find the distance traveled? A little reflection will lead us to the conclusion that distance and speed stand in the same relationship as speed

4. The Integral

and acceleration - that is for constant speed v the distance traveled in time ΔT is $x = v * \Delta T$, just as for constant acceleration a the speed obtained in time ΔT is $v = a * \Delta T$. With this insight, and assuming that we know the distance at some time T_1, we can get the distance at T_2 by starting with $x(T_1)$ and adding the effect of the velocity (here taken as a function of time, $v(t)$) from T_1 to T_2:

$$x(T_2) = x(T_1) + \int_{T_1}^{T_2} v(t) * dt.$$

We can put all this together as follows. Assume that we are interested in what happens from T_1 to T_2, starting with an object at $x(T_1)$, with initial speed of $v(T_1)$, and subject to constant acceleration a_0. Then the velocity at T_2 is

$$v(T_2) = v(T_1) + \int_{T_1}^{T_2} a_o * d\tau = v(T_1) + a_o * (T_2 - T_1)$$

(note that τ is a variable of integration and the result does not depend on τ).

To get the distance traveled we need the speed at any time t, which is simply the above expression with the end time replaced with an arbitrary time t:

$$v(t) = v(T_1) + \int_{T_1}^{t} a_o * d\tau = v(T_1) + a_o * (t - T_1).$$

Given $v(t)$ we can now find the distance at time t as

$$x(t) = x(T_1) + \int_{T_1}^{t} v(\tau) * d\tau = x(T_1) + \int_{T_1}^{t} [v(T_1) + a_o * (\tau - T_1)] * d\tau$$

$$= x(T_1) + [v(T_1) - a_o * T_1] * \int_{T_1}^t d\tau + a_o * \int_{T_1}^t \tau * d\tau$$

$$= x(T_1) + [v(T_1) - a_o * T_1] * (t - T_1) + a_o * (\frac{t^2}{2} - \frac{T_1^2}{2})$$

or, finally,

$$x(t) = x(T_1) + v(T_1) * (t - T_1) + \frac{a_o * T_1^2}{2} + \frac{a_o * t^2}{2} \quad (4.2)$$

In the simple case where $T_1 = 0$, $x(T_1) = 0$, and $v(T_1) = 0$ this simplifies to

$$x(t) = \frac{a_o * t^2}{2}.$$

That is what Motivational Question 2.12 was asking.

4.4. Motivational Problem 2.11 Solution

Motivational Problem 2.11 asks us to find the angle that a cannon should be shot at in order to maximize the distance traveled by the cannon ball. As stated, the cannon is located on a flat plain and the speed, size, and mass of each shot is consistent.

Figure 4.3 illustrates the initial situation when the cannon is first fired. We note that the velocity (remember that

4. The Integral

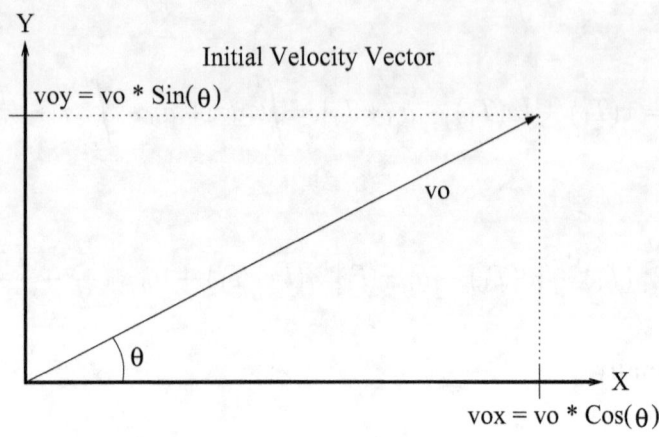

Figure 4.3.: Initial Velocity Vector

velocity is a vector that describes both the speed and the direction of travel) of the cannon ball has two parts - the velocity in the vertical direction and the velocity in the horizontal direction. The initial length of the vector is v_o.

What will be the acceleration experienced by the cannon ball? Gravity acts in the vertical direction, exerting a (assumed) constant acceleration of $-G$ where the negative sign reminds us that the vertical speed will be continuously reduced by the acceleration. Note that there is no acceleration on the horizontal component of the velocity, so we expect the horizontal speed to remain constant.

If we use MKS units, we can measure position in meters (m), speed in meters per second (m/sec), and acceleration in meters per second per second (m/sec^2). The acceleration of gravity is about 10 m/sec^2.

With this as background, an outline of a solution can be seen: the cannon ball initially will have an upward (positive) component of velocity, but that component will be slowed by gravity until its velocity is reversed and it returns to earth. All the time that the cannon ball is above the earth, it will be traveling at a constant horizontal velocity. If we can find the time that it takes for the ball to return to earth, say T_F, we can easily find how far it traveled horizontally during that time.

We can express the velocity of the cannon ball as

$$\vec{v} = v_x * \hat{\epsilon}_x + v_y * \hat{\epsilon}_y$$

where v_x and v_y are the x and y magnitudes of the velocity and $\hat{\epsilon}_x$ and $\hat{\epsilon}_y$ are the x and y unit vectors.

At $t = 0$ the cannon is fired and the initial velocity is

$$\vec{v_o} = v_{ox} * \hat{\epsilon}_x + v_{oy} * \hat{\epsilon}_y$$

where v_{ox} and v_{oy} are the x and y magnitudes of the velocity at $t = 0$. If the initial total speed is v_o we must have

$$(v_{ox}^2 + v_{oy}^2)^{1/2} = v_o$$

and

$$v_{ox} = v_o * \cos(\theta),$$

$$v_{oy} = v_o * \sin(\theta).$$

Consider first the y component of speed:

$$v_y = v_{oy} - G * t$$

which simply states that the initial vertical speed decreases due to the acceleration, towards the earth, of gravity. The x

component is even simpler, since, as noted above, the speed is constant:

$$v_x = v_{ox}$$

Given the two components of the velocity, and taking the position of the cannon as the origin at $(0,0)$, using equation 4.2 we can write down the position of the ball at any time as

$$x(t) = v_{ox} * t$$

and

$$y(t) = v_{oy} * t - G * t^2/2$$

or, as a vector

$$\vec{p}(t) = v_{ox} * t * \hat{\epsilon}_x + (v_{oy} * t - G * t^2/2) * \hat{\epsilon}_y.$$

The time of flight noted above, T_F, is now easily found - it is the time at which the y position of the ball equals 0:

$$v_{oy} * T_f - G * T_f^2/2 = 0$$

which can be solved using the quadratic equation. Rearranging to standard form we get:

$$G * T_F^2/2 - v_{oy} * T_F = 0$$

so that

$$T_F = (v_{oy} \pm \sqrt{v_{oy}^2})/G.$$

The solution we want is $T_F = 2 * v_{oy}/G$. The second solution, $T_F = 0$, is also valid, being the initial time when the ball is at the ground!

Given T_F we can write down the x distance that the ball travels before returning to the ground, say x_F, as simply $v_{ox} * T_F$ or

$$x_F = v_{ox} * 2 * v_{oy}/G$$

$$= v_o * \cos(\theta) * 2 * v_o * \sin(\theta)/G$$

$$= 2 * v_o^2 * \cos(\theta) * \sin(\theta)/G. \qquad (4.3)$$

The problem asks for the maximum distance that the ball can be shot to by varying θ so we look for the maximum of x_F. Note that we only need to consider θ between 0 and 90 degrees (or $\pi/2$ radians). Taking the derivative and setting it equal to 0 gives:

$$d[\sin(\theta) * \cos(\theta)]/d\theta = \sin(\theta) * (-\sin(\theta) + \cos(\theta) * \cos(\theta)$$

$$= \cos^2(\theta) - \sin^2(\theta) = 0$$

or

$$\cos^2(\theta) = \sin^2(\theta).$$

Using equations 1.6 and 1.7 this can be written as

$$\frac{1}{2}[1 + \cos(2\theta)] = \frac{1}{2}[1 - \cos(2\theta)]$$

or

$$\cos(2\theta) = -\cos(2\theta)$$

For θ between 0 and $\pi/2$ the only solution is $2\theta = \pi/2$ so the angle that gives us the longest shot is

$$\theta_{max} = \pi/4 = 45^o.$$

4. The Integral

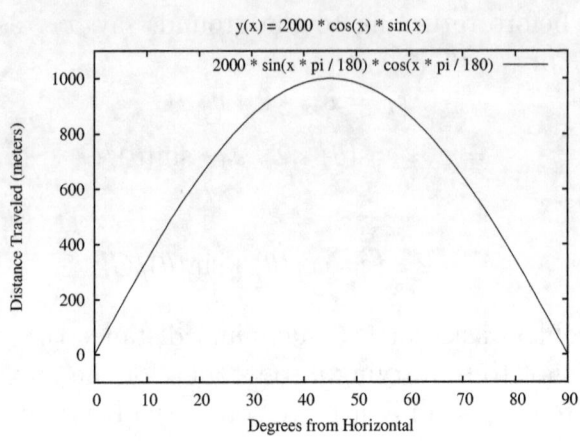

Figure 4.4.: x_F versus θ

Figure 4.4 shows a plot of Equation 4.3 for $v_o = 100m/sec$ and $G = 10m/sec^2$. As expected, the maximum distance traveled occurs at $\theta_{max} = \pi/4 = 45^o$.

The longest shot is therefore

$$x_F = 2 * v_o^2 * \frac{1}{\sqrt{2}} * \frac{1}{\sqrt{2}} * /G$$

$$= v_o^2/G$$

and occurs at $\theta = \pi/4$. Any other angle results in a shorter distance along the x axis.

For example, with $v_o = 100m/sec$ and $G = 10m/sec^2$, $x_F = 1Km$.

Figure 4.5 shows the calculated trajectories of the cannon ball for $\theta = 10^o, 22.5^o, 45^o, 67.5^o$, and 80°.

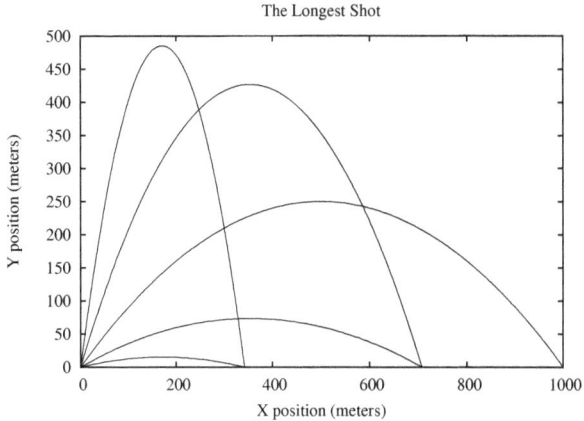

Figure 4.5.: The Longest Shot

4.5. Motivational Problem 2.12 - Solution

We are asked to find the distance traveled by a car whose speed is shown in Figure 4.8.

To see how to solve this, think for a moment about a car that travels at a constant speed, say v_o. We know that the distance traveled is just

$$d = v_o * t$$

where d is the distance traveled in time t.

If we plot the distance traveled versus time we get a straight line with slope v_o as shown in Figure 4.7.

Another way to look at this situation (of constant speed) is to plot the speed versus time as shown in Figure 4.8.

97

4. The Integral

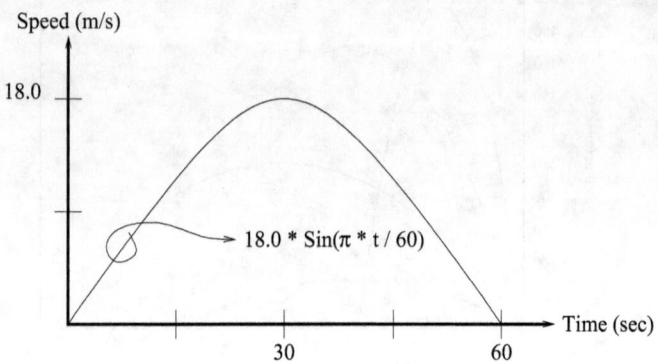

Figure 4.6.: Speed vs. Time Plot

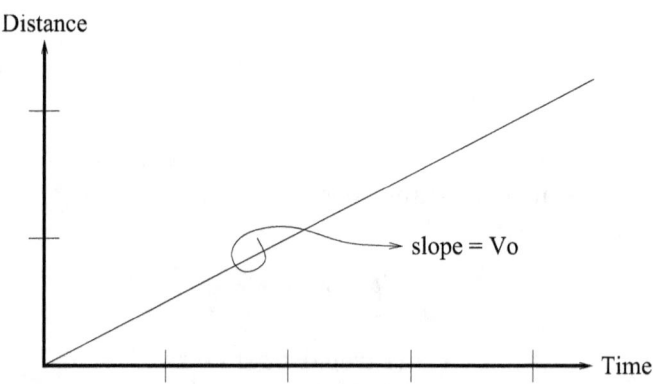

Figure 4.7.: Distance vs. Time Plot - Constant Speed

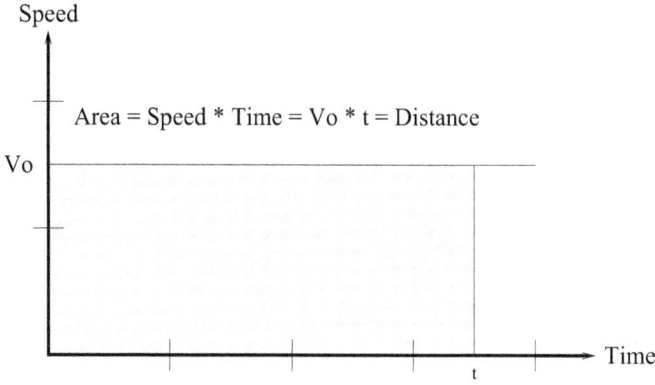

Figure 4.8.: Speed vs. Time Plot - Constant Speed

Thinking about this, we can see that at any time t the car will have traveled a distance given by the product of v_o and t or the *area under the curve* of speed versus time. This gives us a way to think about the distance traveled by a car traveling at a variable speed - we need to find the area under the speed versus time curve for the car.

We can solve MP 2.12 in two ways - using a Riemann Sum (tedious) or using the Fundamental Theorem of Calculus (better!).

4.5.1. Using Riemann Sum

In Figure 4.9 we show the function $y(t) = 18.0 * \sin(\pi * t / 60)$ and we have the times t at which we want to evaluate $y(t)$, namely $t = \Delta t/2, 3 * \Delta t/2, 5 * \Delta t/2, ...$

Taking all this together, for 8 intervals we have that:

$$Area \simeq \Delta t * 18.0 * \sum_{i=0}^{i=7} \sin(\pi * (\Delta t/2 + i * \Delta t)/60).$$

4. The Integral

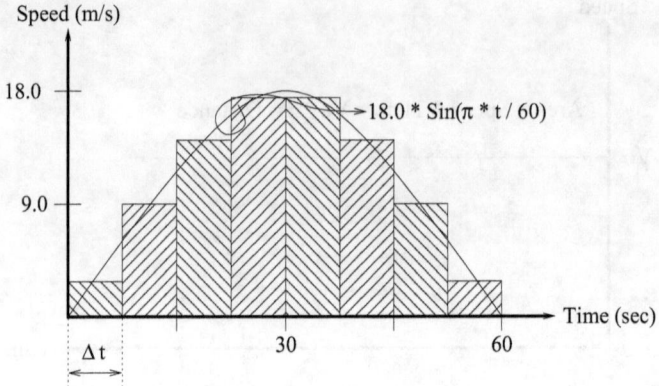

Figure 4.9.: Speed vs. Time

Taking advantage of the symmetry around 30 seconds, recalling that $\Delta t = 60/8$, and simplifying, we can write:

$$Area \simeq 2 * 135 * \sum_{i=0}^{i=3} \sin(\pi * (3.75 + i * 7.5)/60)$$

or

$$Area \simeq 270 * \sum_{i=0}^{i=3} \sin(\pi * (0.0625 + i * 0.125)).$$

Evaluating the four terms in the sum we get

$$Area \simeq 270 * (0.19509 + 0.55557 + 0.83147 + .98079)$$

$$Area \simeq 692.$$

What does this number, 692, represent? It is an approximation of the area under the speed versus time curve, Figure

2.7. The units for this number are therefore speed (measured here in $m/\sec$) times time (measured here in seconds) or the *distance* traveled (in this case in meters).

So, to recap, if a car follows the variable speed curve shown in Figure 2.7 is will travel approximately 692 meters in the indicated 60 seconds. We found this by *approximating* the area under the speed versus time curve using a Riemann Sum.

4.5.2. Using the Fundamental Theorem of Calculus

We really want to find a method of doing integration that does not involve a Riemann Sum. Consider the integration we have been looking at:

$$Area = \int_0^{60} y(t)dt$$

The FFToC states that

$$d(Area(t))/dt = y(t)$$

or

$$d(Area(t))/dt = 18.0 * \sin(\pi * t/60).$$

So, apparently $Area(t)$ is given by a function whose *derivative* is $18.0 * \sin(\pi * t/60)$. We need the corresponding antiderivative to be able to use the FToC.

Recalling that

4. The Integral

$$d(a * \cos(b * x))/dx = -a * b * \sin(b * x)$$

we can write

$$Area(t) = -(18.0 * 60/\pi)\cos(\pi * t/60) + c.$$

where c is an arbitrary constant. Why do we add this constant? Note that all the FFToC states is that the *derivative* of our area function $Area(t)$ is equal to $y(t)$ and this will be true for any arbitrary c. Something must be missing here - what value of x do we use to get our area? Now we use the SFToC! The SFToC states that if

$$\frac{d(Area(t))}{dt} = y(t)$$

then the area under $y(t)$ between $t = a$ and $t = b$ is given as:

$$\int_a^b y(t)dt = Area(b) - Area(a).$$

Now we can complete Motivational Problem 2.12. In terms of the SFToC we have:

$$F(t) = -(18.0 * 60/\pi)\cos(\pi * t/60) + C$$

where $a = 0$ and $b = 60$. Putting this all together we can finally write the *Area* as:

$$Area = 18.0 * \int_0^{60} \sin(\pi * t/60)dt$$

$$= \left[-(18.0 * 60/\pi)\cos(\pi * t/60) + C\right]\big|_{t=60} -$$
$$\left[-(18.0 * 60/\pi)\cos(\pi * t/60) + C\right]\big|_{t=0}$$
$$= 687.55$$

This is to be compared to the estimate obtained using the Riemann Sum. Why do you think the Riemann Sum estimate is a little high?

4.6. Motivational Problem 2.13 - Solution

Motivational problem 2.13 asked us to find the area of a circle. With the idea of an integral as a way of summing up an infinite number of infinitesimal elements, consider figure 4.10. Following [18] the figure shows a circle at the origin with a radius of R. Inside the circle, at a distance r from the center, is a small ring shown shaded. The ring is Δr in width and at the distance r the ring circumference is $2\pi r$. If we cut the ring and straighten it out so that it becomes a rectangle the area of the small ring can be seen to be approximately $2\pi r * \Delta r$. The approximation improves as $\Delta r \to 0$.

If we imagine a series of non-overlapping rings that fill the circle it is clear that the sum of the areas of the rings will equal that of the circle. Each circle, at a radius r, will have an approximate area of the circumference $(2\pi r)$ times Δr or $2\pi r * \Delta r$. The approximation gets better as Δr gets small. If we take N such rings then $\Delta r = R/N$. Taking a Riemann sum to approximate the area results in:

$$Area \simeq \sum_{i=0}^{i=N-1} 2\pi * r_i * \Delta r \qquad (4.4)$$

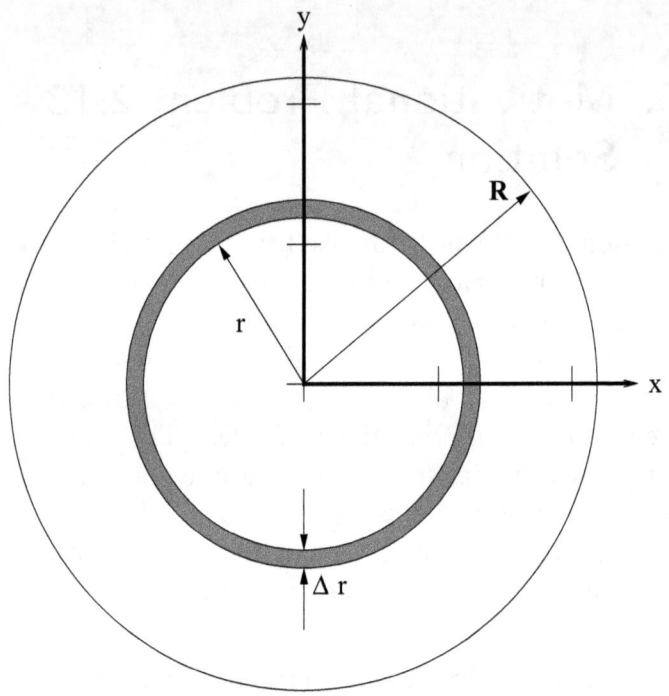

Figure 4.10.: Area of a Circle

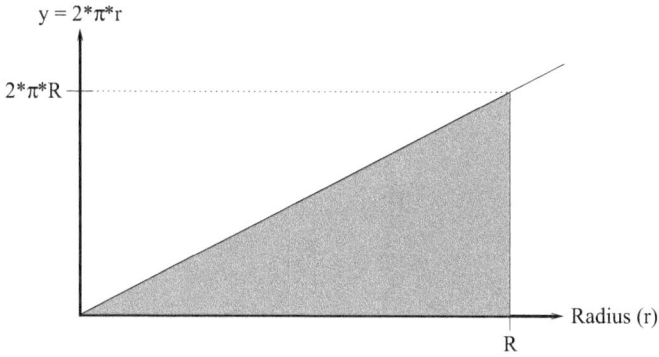

Figure 4.11.: Area of Circle as Area Under a Curve

where r_i is the radius of the i'th ring. We know that a Riemann sum becomes a definite integral as we take the limit as $N \to \infty$ or, equivalently, as $\Delta r \to 0$. Remember a Riemann Sum represents the area under a curve. Looking at Equation 4.4 the curve we are dealing with here is simply $2\pi * r$, and as i ranges from 0 to $N - 1$ the variable r ranges from 0 to R. Figure 4.11 shows the area we are looking for.

The area of the indicated rectangle is $(2 * \pi * R) * R$ so the shaded area, the area of the circle, is $\pi * R^2$ as we expected. Of course, we really wanted to do this using calculus! Writing the integral that represents this area is straightforward - it is:

$$Area = \int_0^R 2\pi r * dr$$

$$= 2\pi \int_0^R r * dr$$

To evaluate the definite integral we need the anti-derivative of r which is known to be $r^2/2$ (see Appendix A). Now we can use the SFToC:

$$Area = 2\pi * [r^2/2] \mid_{r=0}^{r=R}$$

or, finally,

$$Area = \pi R^2.$$

4.7. Simple Integrals

We can apply the SFToC pretty easily to many practical problems. As long as we can get the anti-derivative ($F(x)$) of the function that is being integrated ($f(x)$) we just calculate $F(b) - F(a)$. Appendix A lists some derivative - anti-derivative pairs.

We have already solved a simple integral using a derivative - anti-derivative in Section 4.5.2. As another example, consider finding the area under the curve

$$y(x) = 6 * x^2$$

between $x = 0$ and $x = 5$. Using Appendix A we can write

$$6 * \int_0^5 x^2 dx = [6 * x^3/3] \mid_{x=0}^{x=5} = 250$$

Note that the derivative - anti-derivative pair we used above, namely $\int x^i dx/i = x(i+1)/(i+1) + c$ breaks down when $i = -1$. This case can be handled by the results of Section 3.10 where we found that

$$d(ln(x))/dx = 1/x$$

so that $\int dx/x = ln(x) + c$.

4.8. Motivational Problem 2.14 - Solution

We are now in a position to find the potential energy of a mass above the surface of the earth. If we restrict the height, h, to a small value the potential energy of a mass m can be approximated by

$$U \simeq mgh$$

where g is the acceleration due to gravity. At any height, a small increment of energy can be seen to be

$$\triangle U = mg(h)\triangle h$$

where the notation makes note of the fact that g is not constant but is a function of the height. From physics we know that

$$f = G * M * m/r^2$$

where f is the force of gravitational attraction between masses M (here, the mass of the earth) and m, G is the universal gravitational constant, and r is the distance between the two masses. Here $r = R+h$ where R is the radius of the earth. Recall that acceleration is force (f) divided by mass (m) so we can write the acceleration experienced by m as

$$g(h) = G * M/(R+h)^2$$
$$= K/(R+h)^2$$

Now we can express the increment of energy as

$$\triangle U = m * G * M * \triangle h/(R+h)^2$$

and if we want to get the gravitation potential energy at an arbitrary height we need to evaluate

$$U = m * G * M \int_0^h \frac{dy}{(R+y)^2}$$

From Appendix A we know that

$$\int \frac{dx}{x^2} = \frac{-1}{x} + C$$

Here we can let $(R + y) = x$ from which $dx = dy$. Making this substitution in our expression for U puts it in a form we can readily integrate. Note that, since we are now integrating over the variable x rather than the variable y we also need to map the limits in our expression for U. When $y = o$ note that $x = R$ and when $y = h$ note that $x = R+h$. Re-writing the expression for U in terms of x and its limits gives us

$$U = m * G * M \int_R^{R+h} \frac{dx}{x^2} = m * G * M * [\frac{1}{R} - \frac{1}{(R+h)}]$$

$$= \frac{m * G * M * h}{R * (R+h)} \qquad (4.5)$$

Motivational Question 2.14 also asked us to find the escape velocity from the surface of the earth. We can find the escape velocity by realizing that if we launch a mass from the earth with the escape velocity (V_e) all of its initial kinetic energy $(\frac{1}{2} * m * V_e^2)$ will be converted to gravitational potential energy when the mass comes to rest an infinite distance from the earth. From Equation 4.5, the potential energy of our mass at $h = \infty$ is simply $m * M * G/R$ so if we equate

$$\frac{1}{2} * m * V^2 = m * M * G/R$$

we can find the escape velocity as

$$V = \sqrt{2 * M * G/R}.$$

The earth's mass (M) is $5.9742 \times 10^{24} Kg$, its radius (R) is $6378.1 Km$, and the universal gravitational constant (G) is $6.67300 \times 10^{-11} m^3 Kg^{-1} s^{-2}$. Putting this all together we get the escape velocity of the earth as $11.18 Km/s$ or about $7 miles/s$.

4.9. Integration using Substitution

An integrand may not have an obvious anti-derivative. When this happens, substitution can often be used to simplify a difficult integrand so that an anti-derivative can be found. We actually used substitution in section 4.8 when we substituted $(R + y) = x$. We can further illustrate the power of Substitution with more examples.

4.9.1. Example: $\int (a * x + b) dx$

Let's try to integrate

$$\int (a * x + b) dx.$$

4. The Integral

Start by substituting:

$$u(x) = (a * x + b)$$

and note that

$$u' = d(u(x))/dx = a$$

so that

$$dx = du/a$$

Substituting into the integral then yields

$$\int (a * x + b)dx = (1/a) \int u\,du = (1/a) * (u^2/2) + c.$$

Finally we reverse the substitution to obtain

$$\int (a * x + b) = (1/a) * [(a * x + b)^2/2] + c$$

$$= (a * x^2/2) + b * x + b^2/2a + c$$

Since integration is linear, this problem can be solved without using substitution. Note that

$$\int (a * x + b)dx = \int (a * x)dx + \int b\,dx$$

$$= (a * x^2/2) + b * x + d$$

which is the same answer obtained above within the arbitrary additive constant.

4.9.2. Example: $\int \sin(a * x + b)dx$

Substitute $u(x) = (a * x + b)$ and note that $u' = a$ so that $dx = du/a$. The integration is therefore transformed to:

$$\int \sin(a * x + b)dx = (1/a) \int \sin(u)du$$

$$= -(1/a)\cos(u) + c.$$

Reversing the substitution yields

$$= -(1/a)\cos(a * x + b) + c.$$

4.9.3. Example: $\int_0^\pi \sin(a * x + b)dx$

We can approach this definite integral in the same way as the example above, that is by substituting $u(x) = (a * x + b)$. Again, $u' = a$ so that $dx = du/a$. The integration is therefore transformed to:

$$\int_0^\pi \sin(a * x + b)dx = (1/a)\int_0^\pi \sin(u)du$$

$$= -(1/a)\cos(u) \, |_0^\pi .$$

Reversing the substitution yields

$$= -(1/a)\{\cos(a * \pi + b) - \cos(b)\}.$$

4.9.4. Example: $\int \frac{dx}{(a+x)}$

Let's try a similar substitution as we used above by letting $u = a + x$ so that $u' = 1$ or $du = dx$. Then

$$\int \frac{dx}{(a+x)} = \int \frac{du}{u}.$$

From Appendix A we can write

$$\int \frac{1}{u}du = ln(b * u) + c \qquad (4.6)$$

Reversing the substitution gives

$$\int \frac{dx}{(a+x)} = ln\{b * (a+x)\} + c$$

or, since the constant b in Equation 4.6 can be incorporated into the arbitrary constant,

$$\int \frac{dx}{(a+x)} = ln(a+x) + C.$$

4.9.5. Example: $\int \frac{dx}{(a+x)^2}$

Again let $u = a + x$ so that $dx = du$. Then

$$\int \frac{dx}{(a+x)^2} = \int \frac{du}{u^2} = \int u^{-2}du = c - 1/u = c - 1/(a+x).$$

4.9.6. Example: $\int \ln(a * x + b)dx$

Let $u = a * x + b$ from which we have $du = a * dx$. Making the substitution gives us

$$\int \ln(a * x + b)dx = \frac{1}{a} * \int \ln(u)du$$

$$= \frac{1}{a} * (u)[\ln(u) - 1].$$

$$= \frac{1}{a} * (a * x + b)[\ln(a * x + b) - 1].$$

4.10. Integration by Parts

A second technique of dealing with more complex integrals is to use integration by parts. Consider:

$$\int (f * g')dx.$$

We can use the product rule (Section 3.15) to re-write this as:

$$\int \frac{d(f(x) * g(x))}{dx}dx = \int f(x) * d(g(x))/dx+$$

$$\int g(x) * d(f(x))/dx$$

or, noting that $\int (d(f * g)/dx)dx = f * g$

$$\int (f * g')dx = f * g - \int (g * f')dx$$

So, given a function that we can interpret as the product of f and g' we can recast the required integration from $\int(f * g')dx$ to $\int(g * f')dx$. The second integral may be easier to evaluate than the first.

> Worth Remembering
> Integration by Parts
>
> $$\int f(x) * d(g(x)/dx)dx =$$
>
> $$f(x) * g(x) - \int g(x) * d(f(x)/dx)dx$$
>
> or
>
> $$\int f * g'dx = f * g - \int g * f'dx$$

4.10.1. Example: $\int x * \sin(a * x)dx$

As an example, consider

$$\int x * sin(a * x)dx$$

Let $g' = sin(a * x)$ and $f = x$ so that we can re-write the integral, recalling that an anti-derivative of $g' = sin(a * x)$ is equal to $g = -\cos(a * x)/a$, as

$$\int x * \sin(a * x)dx = -x * \cos(a * x)/a + \int (\cos(a * x)/a)dx$$

$$= -x * \cos(a * x)/a + \sin(a * x)/a^2.$$

This example could also be approached by letting $f = sin(a * x)$ and $g' = x$ so that the integral is re-written as

$$\int x * \sin(a * x)dx = (x^2/2) * \sin(a * x) -$$

$$\int (x^2/2) * a * cos(a * x)dx.$$

Ugh! Although not wrong, clearly this decomposition did not help, resulting in an even more complicated integral to evaluate.

This example illustrates a basic strategy to follow when applying integration by parts - namely to pick f to be a function that is easily differentiated (and has a simple derivative) and to pick g' as one that is easily integrated (and has a simple anti-derivative). With some practise, integration by parts can be a powerful tool.

4.10.2. Example: $\int \log_b(a * x)dx$

Here is another example. Let's find

$$\int \log_b(a * x)dx.$$

Note that $\log(a * x) = \log(a) + \log(x)$ so for now let's focus on

$$\int \log_b(x)dx$$

Since we know the derivative of $\log_b(x)$ and the integral of 1 is trivial, we take

$$f = \log_b(x)$$

$$g' = 1$$

so that

$$f' = (1/x) * \log_b(e)$$

$$g = x + C.$$

Using integration by parts we get

$$\int \log_b(x)dx = \log_b(x) * (x+C) - \log_b(e) * \int (x+C) * (1/x)dx$$

$$= x * \log_b(x) + C * \log_b(x) - \log_b(e) \int dx - C * \log_b(e) * \int (1/x)dx$$

Note that

$$C * \log_b(e) * \int (1/x)dx = C * \log_b(e) * \ln(x) + d$$

4. The Integral

$$= C * \log_b(x) + d$$

so we obtain

$$\int \log_b(x)dx = x * \log_b(x) - x * \log_b(e) + D.$$

Adding the previously ignored term $\int \log_b(a)dx = \log_b(a)*x$ we end up with

$$\int \log_b(a * x)dx = x * \log_b(a * x) - x * \log_b(e) + c.$$

If $a = 1$ and we use the natural logarithms we get

$$\int \ln(x)dx = x * \ln(x) - x + c.$$

A. Collection of Integrals, Functions, and Derivatives

$y(x) \mid \int y(x)dx$	$d(y(x))/dx \mid y(x)$
$c + a * x$	a
$c + a * x^2/2$	$a * x$
$c + a * x^i$	$a * i * x^{i-1}$
$c + a * \log_b(d * x)$	$a * (1/x) * \log_b(e)$
$c + a * x * \log_b(x/e)$	$a * \log_b(d * x)$
$a * \ln(b * x)$	$a * (1/x)$
$c + a * x * [\ln(x) - 1]$	$a * \ln(d * x)$
$c + \ln(a * x + b)$	$a/(a * x + b)$
$c + (a * x + b)[\ln(a * x + b) - 1]/a$	$\ln(a * x + b)$
$c + a * e^{b*x}$	$a * b * e^{b*x}$
$c + a * \sin(b * x)$	$a * b * \cos(b * x)$
$c + a * \cos(b * x)$	$-a * b * \sin(b * x)$

A. Collection of Functions

Note that the columns can be interpreted in two ways:

Left Column $= y(x)$ | Right Column $= d(y(x))/dx$

or

Left Column $= \int y(x)dx$ | Right Column $= y(x)$

In this table a, b, d are constants and c is the arbitrary constant that must be added to any indefinite integral. Note that the $\int y(x)dx$ column is the anti-derivative of the $y(x)$.

Index

acceleration, 25
Antiderivative, 83

cannon, 25
capacitor, 19
Chain Rule, 67
composite function, 67
compound interest, 9
cost curve, 22

Definite Integral, 85
demand curve, 22
Derivative, 35
    of a Polynomial, 41
    of the Cos(x), 40
        using Series Expansion, 45
    of the Exponential, 46
        using Series Expansion, 46
    of the Logarithm, 52
    of the Sin(x), 40
        using Series Expansion, 44
Derivatives
    using Series Expansions, 44

equations of motion, 87
Euler's Identity, 63

GnuPlot, 12

Integral, 79
Integration
    by Parts, 113
    using Substitution, 109
isotope, 21

L'Hôpital's Rule, 71
Linearity of Differentiation, 38

marginal cost, 59
marginal revenue, 59
Maximum Value, 54
Minimum Value, 54
Motivational Problem
    Acceleration, Speed, Distance, 25
    Age of the Earth, 21
    Area of a Circle, 27
    Designing Pastures, 21
    Distance Traveled, 26
    Euler's Identity, 22
    Maximizing Profit, 22

Maximum Package Volume, 24
Potential Energy of Gravity, 27
Slope of a Curve, 17
The Capacitor, 19
The Longest Shot, 25
Multiple Angle Formulas, 14

Ohm's Law, 20, 49

Pastures, 21
Product Rule, 60

Quotient Rule, 63

Radioactive Decay, 21
Riemann Sum, 79
Rule
    Chain, 67
    Product, 60
    Quotient, 63

Simple Integrals, 106
slope - intercept, 17
Slope of a Curve, 17, 18
speed, 25

Theorem of Calculus
    First Fundamental, 81
    Second Fundamental, 83
time constant, 50

uranium, 21

Bibliography

[1] bc - an arbitrary precision numeric processing language
http://www.gnu.org/software/bc/

[2] The Gnuplot Home Page, http://www.gnuplot.info/

[3] Equations and Graphs of Straight Lines, Math 1010 On-Line, University of Utah, http://www.math.utah.edu/online/1010/straight/

[4] Wikipedia Article - Leibniz Notation http://en.wikipedia.org/wiki/Leibniz_notation

[5] Tung, K. K., Topics in Mathematical Modeling, Princeton:Princeton, c2007, p. 69.

[6] Wikipedia Article - Linear http://en.wikipedia.org/wiki/Linear

[7] Wikipedia Article - Binomial Expansion http://en.wikipedia.org/wiki/Binomial_expansion

[8] Wikipedia Article - Trigonometric Function http://en.wikipedia.org/wiki/Trigonometric_function

[9] Wikipedia Article - Exponential Function http://en.wikipedia.org/wiki/Exponential_function

Bibliography

[10] Pasles, Paul C., Benjamin Franklin's Numbers, Princeton Press:Princeton, New Jersey, c. 2008, p. 98.

[11] Banner, Adrian, <u>The Calculus Lifesaver</u>, Princeton University Press:Princeton, c2007, p. 174.

[12] Lawrence S. Husch, Visual Calculus, Product Rule, University of Tennessee, Knoxville, Mathematics Department `http://archives.math.utk.edu/visual.calculus/2/product_rule.5/`

[13] Wikipedia Article - Maximizing Profit `http://en.wikipedia.org/wiki/Profit_maximization`

[14] Smith, J.O. in "Mathematics of the Discrete Fourier Transform" `https://ccrma.stanford.edu/~jos/mdft/Euler_s_Identity.html`

[15] Wikipedia Article - Euler's Formula `http://en.wikipedia.org/wiki/Euler's_formula`

[16] Riemann Sums Demonstration - Univeristy of Tennessee at Knoxville `http://archives.math.utk.edu/visual.calculus/4/riemann_sums.4/`

[17] Weisstein, Eric W., <u>CRC Concise Encyclopedia of Mathematics</u>, Chapman & Hall/CRC, Boca Raton, 1999.

[18] Hahn, Karl, Coached Exercise: Alternative Method for the Area of a Circle `http://www.karlscalculus.org/CoachedCircle.html`